Computational Intelligence and Bioengineering

Computational Intelligence and Bioengineering

Contributors

C. Nataraj, A. Jalali et al.

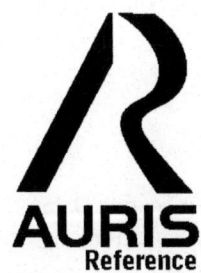

AURIS
Reference

www.aurisreference.com

Computational Intelligence and Bioengineering

Contributors: C. Nataraj, A. Jalali et al.

Published by Auris Reference Limited
www.aurisreference.com

United Kingdom

Chemical Soil Stabilization

ISBN: 978-1-78154-846-2

British Library Cataloguing in Publication Data
A CIP record for this book is available from the British Library

Printed in the United Kingdom
Exclusively distributed by CBS Publishers & Distributors Pvt. Ltd.
Sales & Distribution Rights only for India, Pakistan, Bangladesh, Sri Lanka, Nepal and Bhutan. This book is not to be sold outside these territories.

Contents

List of Abbreviations...vii

List of Contributors..ix

Preface.. xiii

Chapter 1   Application of Computational Intelligence Techniques for
             Cardiovascular Diagnostics... 1

Chapter 2   The Successive Zooming Genetic Algorithm and its Applications........ 37

Chapter 3   Public Portfolio Selection Combining Genetic Algorithms
             and Mathematical Decision Analysis................................... 57

Chapter 4   The Search for Parameters and Solutions: Applying Genetic
             Algorithms on Astronomy and Engineering 83

Chapter 5   Fusion of Visual and Thermal Images Using Genetic Algorithms 115

Chapter 6   Optimal Feature Generation with Genetic Algorithms and FLDR in
             a Restrictedvocabulary Speech Recognition System......................... 147

Chapter 7   Performance of Varying Genetic Algorithm Techniques in Online
             Auction 181

Chapter 8   Modelling the Innate Immune System... 217

Chapter 9   Optimal Design of Power System Controller Using Breeder
             Genetic Algorithm 241

Chapter 10   Performance Study of Cultural Algorithms Based on Genetic
             Algorithm with Single and Multi Population for the MKP 259

             Citations.. 285

             Index... 287

List of Abbreviations

ANN	Artificial Neural Networks
BGA	Breeder Genetic Algorithm
BP	Blood Pressure
CA	Cultural Algorithm
CGA	Continuous Genetic Algorithms
CWT	Continuous Wavelet Transform
DM	Decision Makers
DRC	Dynamic Range Compression
DTW	Dynamic Time Warping
EA	Evolutionary Algorithm
EC	Evolutionary Computation
ESO	European Southern Observatory
FDR	Fisher Discriminant Ratio
FI	Fused Image
FLDR	Fisher's Linear Discriminant Ratio
GA	Genetic Algorithm
HHM	Hidden Markov Models
HIS	Human Immune System
HR	Heart Rate
HVS	Human Visual System
LPC	Linear Predictive Coding
MCKP	Multi Choice Knapsack Problems
MKP	Multiple Knapsack Problems
MLP	Multi-Layered Perceptron
MSR	Multi-Scale Retinex
PCA	Principal Component Analysis
PDE	Partial Differential Equation
PGA	Parallel Genetic Algorithm
PLIF	Pixel-Level Image Fusion
ROC	Receiver Operating Curve
SAC	Self-Adaptive Crossover
SAIFI	System Average Interruption Frequency Index
SED	Spectral Energy Distribution
SSR	Single-Scale Retinex
SVM	Support Vector Machine
WT	Wavelet Transforms

List of Contributors

C. Nataraj, A. Jalali
Department of Mechanical Engineering, Villanova University, Villanova, Pennsylvania, USA

P. Ghorbanian
Department of Mechanical Engineering, Villanova University, Villanova, Pennsylvania, USA

Young-Doo Kwon
School of Mechanical Engineering & IEDT, Kyungpook National University,

Dae-Suep Lee
Division of Mechanical Engineering, Yeungjin College, Daegu, Republic of Korea

Eduardo Fernández-González
Autonomous University of Sinaloa México

Inés Vega-López
Autonomous University of Sinaloa México

Jorge Navarro-Castillo
Autonomous University of Sinaloa México

Annibal Hetem Jr.
Universidade Federal do ABC Brasil

Sertan Erkanli
Turkish Air Force Academy, Turkey
Old Dominion University, USA

Jiang Li
Old Dominion University, USA

Ender Oguslu
Turkish Air Force Academy, Turkey
Old Dominion University, USA

Julio César Martínez-Romo
Instituto Tecnológico de Aguascalientes, Mexico

Francisco Javier Luna-Rosas
Instituto Tecnológico de Aguascalientes, Mexico

Miguel Mora-González
Universidad de Guadalajara, Centro Universitario de los Lagos, Mexico

Carlos Alejandro de Luna-Ortega
Universidad Politécnica de Aguascalientes, Mexico

Valentín López-Rivas
Instituto Tecnológico de Aguascalientes, Mexico

Kim Soon Gan
Universiti Malaysia Sabah, School of Engineering and Information Technology, Sabah Malaysia

Patricia Anthony
Universiti Malaysia Sabah, School of Engineering and Information Technology, Sabah Malaysia

Jason Teo
Universiti Malaysia Sabah, School of Engineering and Information Technology, Sabah Malaysia

Kim On Chin
Universiti Malaysia Sabah, School of Engineering and Information Technology, Sabah Malaysia

Pedro Rocha
Federal University of Juiz de Fora, UFJF Brazil

Alexandre Pigozzo
Federal University of Juiz de Fora, UFJF Brazil

Bárbara Quintela
Federal University of Juiz de Fora, UFJF Brazil

Gilson Macedo
Federal University of Juiz de Fora, UFJF Brazil

Rodrigo Santos
Federal University of Juiz de Fora, UFJF Brazil

Marcelo Lobosco
Federal University of Juiz de Fora, UFJF Brazil

K. A. Folly
University of Cape Town Private Bag., Rondebosch 7701 South Africa

S. P. Sheetekela
University of Cape Town Private Bag., Rondebosch 7701 South Africa

Deam James Azevedo da Silva
Universidade Federal do Pará (UFPA), Brazil

Otávio Noura Teixeira
Centro Universitário do Estado do Pará (CESUPA) Brazil

Roberto Célio Limão de Oliveira
Universidade Federal do Pará (UFPA), Brazil

Preface

Computation intelligence paradigms including artificial neural networks, fuzzy systems, evolutionary computing techniques, intelligent agents and so on provide a basis for human like reasoning in medical systems. Bioengineering is the application of engineering knowledge to the fields of medicine and biology. The text *Computational Intelligence and Bioengineering* presents a collection of recent studies covering the computational intelligence applications with emphasis on their application to challenging real-world problems encountered by humans. First chapter focuses on application of computational intelligence techniques for cardiovascular diagnostics. The successive zooming genetic algorithm and its applications have been presented in second chapter. Public portfolio selection combining genetic algorithms and mathematical decision analysis have been proposed in third chapter. Fourth chapter discusses how to apply genetic algorithms on astronomy and engineering. The goal of fifth chapter is to develop computational methods for obtaining efficiently improved images. In sixth chapter, we discuss a method to implement a high performance, real-time, restricted-vocabulary speech recognition system, combining a genetic algorithm and the Fisher's linear discriminant Ratio (FLDR) in its matrix formulation. In seventh chapter, variations of genetic algorithms are applied in optimizing the bidding strategies for a dynamic online auctions environment. Eighth chapter describes the GPU-based implementation of the sensitivity analysis and also presents some of the sensitivity analysis results. In ninth chapter, breeder genetic algorithm (BGA) with adaptive mutation is used for the optimization of the parameters of the power system stabilizer (PSSs). Last chapter focuses on performance study of cultural algorithms based on genetic algorithm with single and multi-population for the multidimensional knapsack problem (MKP).

Chapter 1

APPLICATION OF COMPUTATIONAL INTELLIGENCE TECHNIQUES FOR CARDIOVASCULAR DIAGNOSTICS

C. Nataraj, A. Jalali and P. Ghorbanian

Department of Mechanical Engineering, Villanova University, Villanova, Pennsylvania, USA

INTRODUCTION

Cardiovascular disease, including heart disease and stroke, remains the leading cause of death around the world. Yet, most heart attacks and strokes could be prevented if it were possible to provide an easy and reliable method of monitoring and diagnostics. In particular, the early detection of abnormalities in the function of the heart, called arrhythmias, could be valuable for clinicians. Hemodynamic instability is most commonly associated with abnormal or unstable blood pressure (BP), especially hypotension, or more broadly associated with inadequate global or regional perfusion. Inadequate perfusion may compromise important organs, such as heart and brain, due to limits on coronary and cerebral auto regulation and cause life-threatening illnesses, or even death. Therefore, it is crucial to identify patients who are likely to become hemodynamically unstable to enable early detection and treatment of these life-threatening conditions (Cao, Eshelman et al. 2008). Modern intensive care units (ICU) employ continuous hemodynamic monitoring (e.g., heart rate (HR) and invasive arterial BP measurements) to track the state of health of the patients. However, clinicians in a busy ICU would be too overwhelmed with the effort required to assimilate and interpret the tremendous volumes of data in order to arrive at working hypotheses. Consequently, it is important to seek to have automated algorithms that can accurately process and classify the large amount of data gathered and to identify patients who are on the verge of becoming unstable (Cao, Eshelman et al. 2008). Modern ICUs are equipped with a large array of alarmed monitors and devices which are used to try to detect clinical changes at the earliest possible moment so as to prevent any further deterioration in a patient's condition. The effectiveness of these systems depends on the sensitivity and specificity of the alarms, as well as on the response of the ICU staff to the alarms. However, when large numbers of alarms are either technically false, or true, but clinically irrelevant, response efficiency can be decreased, reducing the quality of patient care and increased patient (and family) anxiety (Laramee, Lesperance et al. 2006). It is patently obvious that physiological time series such as hemodynamic and electrophysiological data represent the physiological state of subjects in a medical environment.

These time series are collected over long periods of time and are usually a source of a large number of interesting behaviors or features which have the potential to be used in identifying and predicting a subject's current and future state of health. However, the high dimensionalities and complexity of the measured physiological signals make the interpretation and analysis difficult, if not impossible. Hence, although they clearly contain useful information, these signals cannot be used directly. Extraction of such hidden information can be addressed using the concept of feature extraction. Essentially, feature extraction is focused on dimensionality reduction and on revealing information from the different time scales that underlie physical phenomena. Also of importance is the concept of classification, where the features are employed in an intelligent algorithm to classify the patient, for example, as healthy or sick. Clearly, this is a broad area with an increasingly diverse set of applications. In order to illustrate the power and utility of these methods, and given the limited space, we limit ourselves to two examples both of which illustrate feature extraction and classification approaches. The first application discussed in this chapter is the detection of cardiac arrhythmia detection. In this application, we apply continuous wavelet transform (Daubechies 2006) and principal component analysis (Jolliffe 2002) as feature extraction tools and artificial neural network algorithm as a classifier (Caudill 1989). The second application discussed concerns the identification of ICU patients. In this example, we apply some novel feature extraction techniques to highlight the differences between healthy and patient subjects. Then we apply fuzzy decision theory (Zadeh 1968) as a final classifier.

AN IMPROVED PROCEDURE FOR DETECTION OF HEART ARRHYTHMIAS

The electrocardiogram (ECG) plays an important role in the process of monitoring and preventing heart attacks. The typical ECG, shown in Figure 1, consists of three basic waves: P, QRS, and T. These waves correspond to the far field induced by specific electrical phenomena on the cardiac surface, namely, the atrial depolarization, P, the ventricular depolarization, QRS complex, and the ventricular repolarization, T. It should be noted however that the ECG signal does not look the same in all the leads of the standard 12-lead system used in clinical practice. There is increasing recognition that computer-based analysis and classification of diseases could be very helpful in diagnostics and several algorithms have been reported in the literature for detection and classification of ECG beats using artificial neural networks (ANN). It has indeed been shown that neural networks are particularly able to recognize and classify ECG signals more accurately than other classification methods (Ozbay and Karlýk 2001). The techniques, developed for automated detection of changes in electrocardiographic signals, work by transforming the mostly qualitative diagnostic criteria into a more objective quantitative signal feature classification problem. This transformation of the ECG signals has been carried out in the past using techniques such as autocorrelation function, time frequency analysis, and wavelet transforms (WT) (Maglaveras, Stamkopoulos et al. 1998; Addison, Watson et al. 2000; Kundu, Nasipuri et al. 2000; Dokur and Olmez 2001; Saxena, Kumar et al. 2002). Results of these and other studies in the literature have demonstrated that WT is the most promising method to extract features that characterize the behavior of ECG signals in an effective manner.

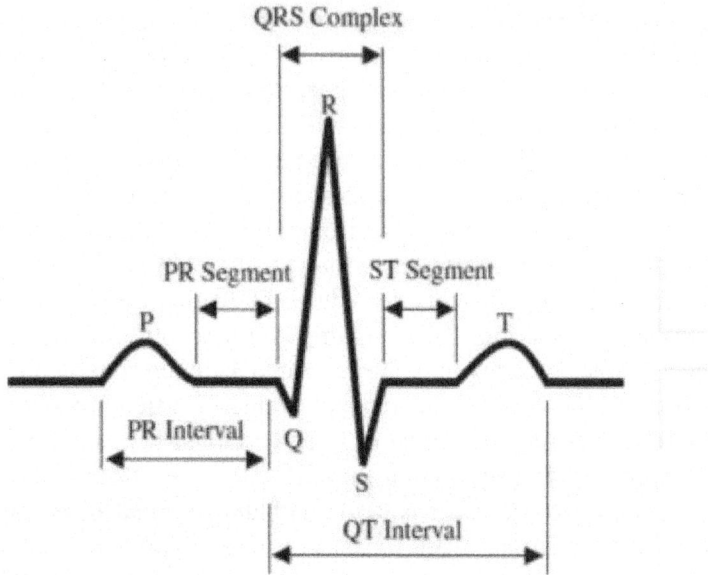

Figure 1: The components of the ECG signal.

A study of the nonlinear dynamics of electrocardiogram signals for arrhythmia characterization was presented by Owis (Owis, Abou-Zied et al. 2002). They selected the correlation dimension and the largest Lyapunov exponent as two features for characterizing five different classes of ECG signals. The statistical analysis of the calculated features indicated that they differ significantly between the normal heart rhythm and the different arrhythmia types and, hence, can be somewhat useful in ECG arrhythmia detection. However, their study is limited by the fact that the discrimination between different arrhythmia types is difficult using those features. Application of the wavelet transform, principal component analysis (PCA) and several types of artificial neural network structures to detect and classify different kinds of heart arrhythmias have also been reported (Silipo and Marchesi 1998); this study compared results of different neural network structures in order to find the best one for the classification of specific types of arrhythmias. A neural network classifier was used by (Christov and Bortolan 2004) to recognize premature ventricular contraction arrhythmia beats in an ECG signal database. A combination of neural network and discrete wavelet transform (DWT) has also been applied for detecting four types of heart arrhythmias (Guler and Ubeyli 2005). Another application of a combination of wavelet transform and ANN in arrhythmia detection is proposed in the study by Vikas (Vikas and Sahambi 2004). In the first step, a set of discrete wavelet transform coefficients which contain the maximum information about the arrhythmia is selected from the wavelet decomposition. Then, these coefficients, in addition to the information about the RR interval, QRS duration, and amplitude of the R-peak, are fed into a multi-layer perceptron algorithm. They reach an overall accuracy of 98% in the classification of 47 patient records. Papaloukas, et al. (Papaloukas, Fotiadis et al. 2002) used a neural network classifier to detect and classify ischemic arrhythmia episodes in the ECG signal. They also used PCA to select and extract features from the ECG signal. Lee (Lee, Park et al. 2005) applied linear discriminant analysis to 17 input features, which were based on wavelet coefficients, to reduce the feature dimension from 17 to 4, for arrhythmia detection. Then, a multi-layer perceptron classifier was applied to detect 6 types of arrhythmia beats from a 4-dimensional input feature. Foo (Foo, Stuart et al. 2002) compared and evaluated different types of multilayer neural network structures as the ECG pattern classifiers and finally settled on a two-layer feed-forward neural network. However, their work is limited to detecting only two types of patterns including normal beats and premature ventricular contractions (PVC). Acharya, et al. (Acharya, Bhat et al. 2003) proposed an algorithm based on a neural network classifier and fuzzy cluster to analyze ECG signals. They compared these two classifiers and reported the fuzzy cluster as a better classifier in comparison with the neural one. They classified 4 types of ECG signals including ischemic cardiomyopathy beat, complete heart block beat, atrial fibrillation beat, and normal beat. Also, Ozbay (Ozbay, Ceylan et al. 2006) proposed a comparative study of the classification accuracy of ECG signals using a wellknown neural network architecture, a multi-layered perceptron (MLP) structure, and a new fuzzy clustering neural network architecture (FCNN) for early diagnosis; They used these two classifiers to classify 10 types of ECG signals. Based on their test results they suggested that a new proposed FCNN architecture can generalize better than ordinary MLP architecture and could also learn better and faster. The advantage of their proposed structure was a result of reduction in the number of segments by grouping similar segments in training data with fuzzy C-means clustering. Zhang (Zhang and Zhang 2005) developed an algorithm for recognizing and classifying four types of ECG signal beats including normal beat, left bundle branch block beat, right bundle branch block beat and premature ventricular contraction PVC beat. They extracted the principal characteristics of the signals by means of the PCA technique and they showed that out of 100 principal components, the first 30 principal components have most of the total energy of the data set and hence used it as the input vector for the classifier. Among different types of classifiers, they used the support vector machine (SVM), which has exhibited very good success compared to other classification methods in complicated problems. A comparison between different classifiers is also presented in their research. A comparison between different structures for heart arrhythmia detection algorithms based on neural network, fuzzy cluster, wavelet transform and principal component analysis, was carried out by Ceylan (Ceylan and Ozbay 2007). Kutlu (Kutlu, Kuntalp et al. 2008) applied a K-nearest neighborhood algorithm for the purpose of classification. They extracted features from the electrocardiograph signals by using higher order statistics. They achieved an accuracy of 97.3% in classifying 5 types of heart arrhythmias. Cvikl (Cvikl and Zemva 2010) designed a fieldprogrammable gate array-based (FPGA) system for ECG signal processing. Their system performs QRS complex detection and beat classification into either normal or PVC. They reached a sensitivity of 92.4% for PVC detection. The most difficult problem faced by today's automatic ECG analysis is the large variation in the morphologies of ECG waveforms, not only of different patients or patient groups but also within the same patient. The ECG waveforms may differ for the same patient to such an extent that they could be unlike each other, and at the same time, alike for different types of beats. This is the main reason that the beat classifiers, which were reviewed in this study, perform well on the training data, while generalizing poorly when presented with the ECG waveforms of different patients (Ozbay, Ceylan et al. 2006). We address this problem of beat classifier performance by using a combination of continuous wavelet transform

(CWT) and principal component analysis in order to prepare a more effective input data for the artificial neural network classifier. Since this would lead to a better input vector structure for the neural network classifier, we expect to obtain a better and more accurate performance of the classifier. Moreover, we propose to use a signal filtering method in order to remove ECG signal baseline wandering which can be further expected to improve classification. This section is not focused on improving the processing techniques such as CWT and PCA or on improving the neural network structure. It is instead focused on designing an innovative algorithm which is a combination of these techniques in order to achieve reasonably accurate classification results in the field of heart arrhythmia detection. Although we address a better classification performance in the field of heart arrhythmia detection, another interesting achievement of this study is that the classifier in this study detects 6 types of ECG signals including the normal signal and 5 types of arrhythmia beats. This quantity of ECG signal types studied here is a much larger number in comparison with other studies in this field. The structure proposed in this section is composed of three sub stages:

(a)   continuous wavelet transform, which provides feature extraction;

(b)   principal component analysis, which performs elimination of inconsiderable features; and finally,

(c)   multilayer perceptron neural network, working as a final classifier. The outline of this section is as follows; a basic definition of CWT is presented in Section 2.1. In Section 2.2 the procedure of computing principal components of a data set is provided. In Section 2.3, the designed algorithm of our study is presented with a detailed explanation. Finally, in Section 2.4, the results of our study are presented.

Continuous wavelet transform

The wavelet transform (WT) provides very general techniques, which can be applied to many tasks in signal processing. Wavelet transform can be thought of as an extension of the classic Fourier transform; the difference is that, instead of working on a single scale (time or frequency), it works on a multi-scale basis and describes the signal's frequency content at given times. This multi-scale feature of the WT allows the decomposition of a signal into a number of scales, each scale representing a particular coarseness of the signal under study. Continuous wavelet transform (CWT) is a time-frequency analysis method which differs from the more traditional short time Fourier transform (STFT) by having a variable window width, which is related to the scale of observation. Another important distinction from the STFT is that the CWT is not limited to using sinusoidal analyzing functions (Osowski and Linh 2001);

a large selection of localized waveforms can be employed as the analyzing function. The wavelet transform of a continuous time signal, x (t), is defined as

$$T(a,b) = \frac{1}{\sqrt{a}} \int_{-\infty}^{+\infty} x(t) \psi^* \left(\frac{t-b}{a} \right) dt$$

where $\psi^*(t)$ is the complex conjugate of the analyzing wavelet function $\psi(t)$, a is the dilation parameter of the wavelet, which is called 'scale', and b is the location parameter of the wavelet (Osowski and Linh 2001).

Principal component analysis

Principal component analysis (PCA) has become a well-established technique for feature extraction and dimensionality reduction. An assumption made for feature extraction and dimensionality reduction by PCA is that most of the information of the observation vectors, with the dimension p, is contained in the subspace spanned by the first m principal axes, where . Therefore, each original data vector can be represented by its principal component vector with dimensionality m (Ceylan and Ozbay 2007). This procedure decreases the data dimensionality without significant loss of information (Addison 2005). Principal components analysis has been used in a wide range of biomedical problems, including the analysis of ECG data (Silipo and Marchesi 1998; Wang and Paliwal 2003; Addison 2005; Ceylan and Ozbay 2007). In order to apply PCA on a data set, X, the following five steps are required (Zhang and Zhang 2005; Ceylan and Ozbay 2007):

1. Subtract the mean value, μ, from each of the data dimensions.
2. Calculate the covariance matrix, S.

$$S = \frac{1}{N} \sum_{i=1}^{N} (x_i - \mu)^T (x_i - \mu)$$

where, , μ is the sample mean, and N is the number of samples.

3. Calculate the eigenvectors and eigenvalues of the covariance matrix.
4. Choose the components and form a feature vector. In general, once the eigenvectors are found from the covariance matrix, the next step is to order them by decreasing order of the magnitude of the eigenvalue. Then the feature vector is constructed by taking the corresponding eigenvectors.

Feature Vector = (eig 1 eig 2 eig 3 ... eig n)

5. Derive the new data set. Once the components (or eigenvectors) have

been chosen and the feature vector is constructed, the final data is constructed by pre-multiplying by the transpose of the feature vector as shown below.

Final Data = Row Feature Vector x Row Data Adjust

where, 'Row Feature Vector' is the transpose of the matrix with the eigenvectors in the columns, 'Row Data Adjust' is the transpose of the mean-adjusted data matrix, and 'Final Data' is the final data set, with data items in columns.

Methodology

A schematic of the designed algorithm in this study is shown in Figure 2. This algorithm consists of three stages: pre-processing, main process and finally, classification of the ECG beats.

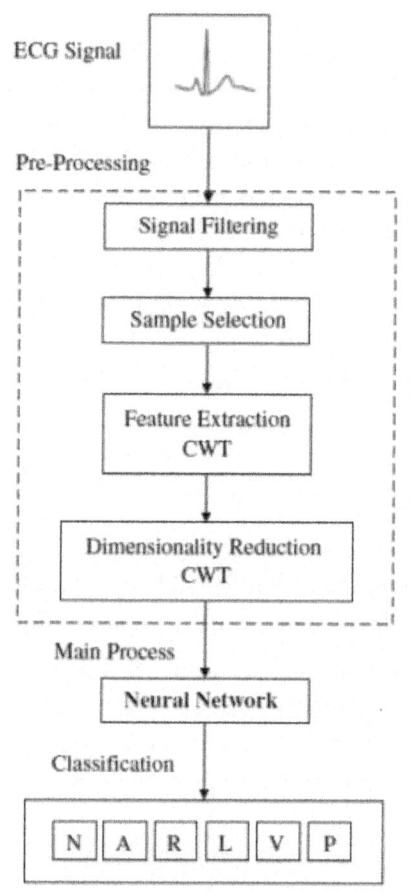

Figure 2: Schematic of the designed algorithm

The data of ECG signals used in this study are taken from the MIT-BIH ECG signal database, including normal beats and five types of different arrhythmia beats. MIT-BIH ECG signal database is a well-known standard database which has been used in many research projects reported in the literature (Silipo and Marchesi 1998; Owis, Abou-Zied et al. 2002; Zhang and Zhang 2005; Ceylan and Ozbay 2007; Cvikl and Zemva 2010). For this study, the selected types of arrhythmias are atrial premature beats (A), right bundle branch block beats (R), left bundle branch block beats (L), paced beats (P), and premature ventricular contraction beats (PVC or V).

Pre-processing

This stage includes four levels of data processing: signal filtering, sample selection, feature extraction, and dimensionality reduction. In the stage of signal filtering, a mathematical method presented by Ghaffari (Ghaffari, SadAbadi et al. 2006) is employed to remove baseline wandering of the ECG signal. Figures 3.a and 3.b show raw ECG signal of records 232 and 208 from the MIT-BIH database, each of which clearly exhibit baseline wandering. Figures 3.c and 3.d show the same ECG signals after applying the filtering method. It is clear that the baseline wandering has been removed, leading to a better performance of the neural classifier. For the stage of sample selection, the suitable range of samples from the raw ECG signal was found experimentally to be 150 samples after the R wave for all types of signals, which together comprise what we call a segment. These segments are found to be an appropriate range of ECG signals which represent morphological differences between different types of ECG beats and include sufficient amount of data needed for classification of heart arrhythmias. For three types of ECG signals under study, the morphologies of ECG beats are shown in Figures 4.a - 6.a; Figures 4.b - 6.b show the selected segments of these beats.

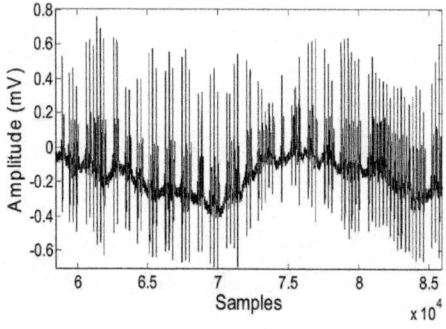

Figure 3a: Raw ECG signal from record 232

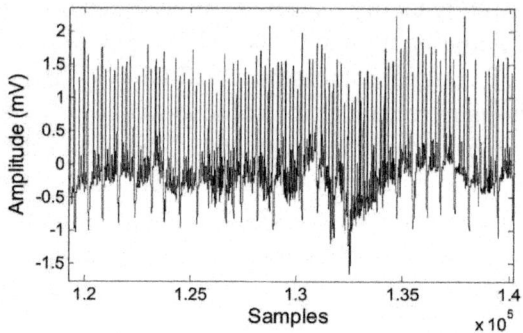

Figure 3b. Raw ECG signal from record 208

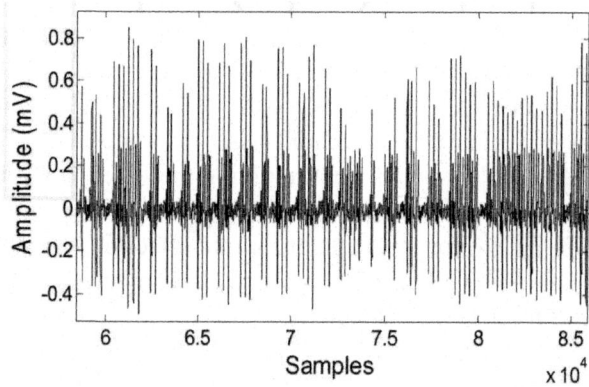

Figure 3c: Filtered ECG signal from record 232

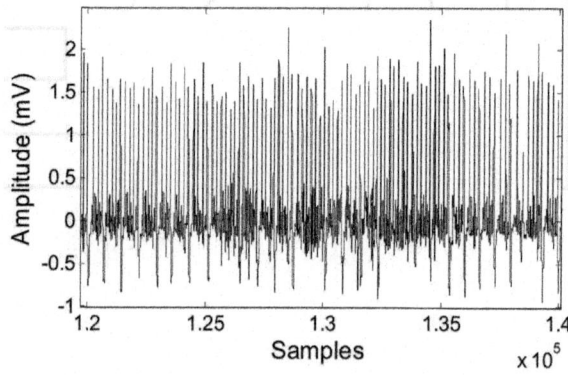

Figure. 3d. Filtered ECG signal from record 208

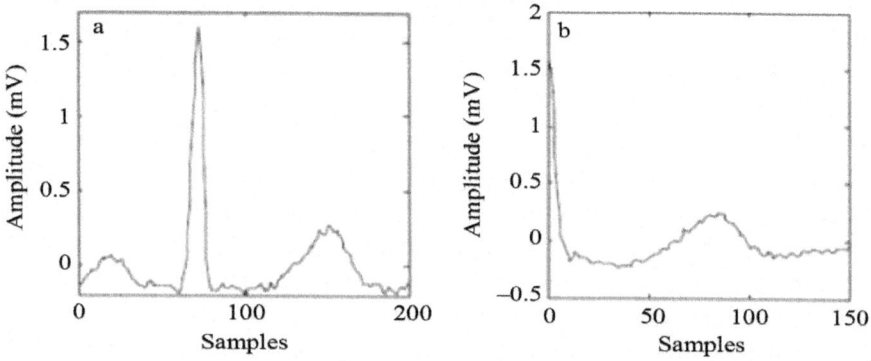

Figure: 4. (a) Normal beat, (b) selected segment for Normal beat

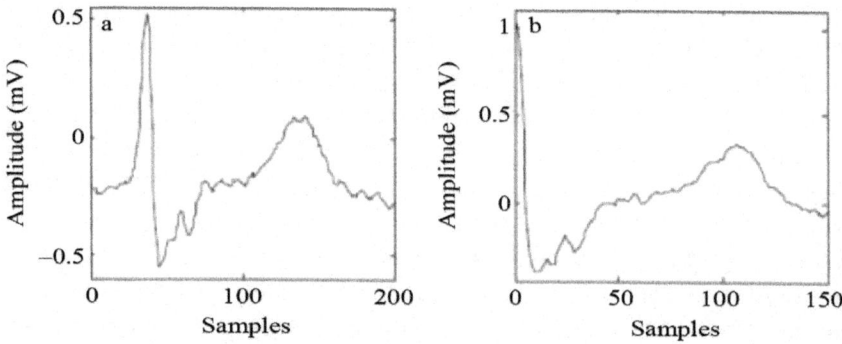

Figure 5: (a) Atrial beat, (b) selected segment for Atrial beat.

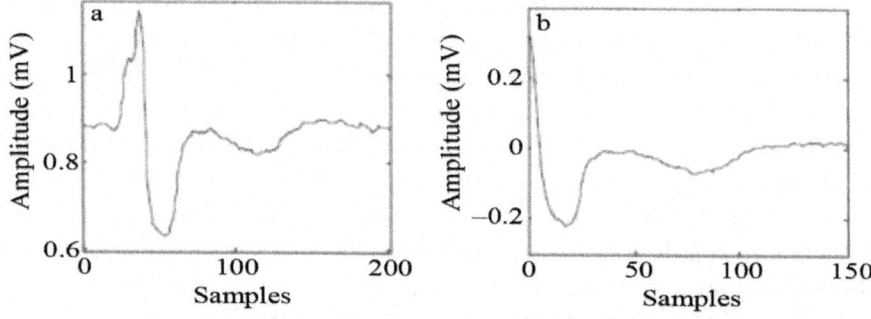

Figure 6: (a) Right Bundle beat, (b) selected segment for Right Bundle beat.

The choice of the analyzing function in wavelet transform, which is called the mother wavelet, has a significant effect on the result of analysis and should be selected carefully based on the nature of the signal (Addison 2005). Several mother wavelets, such as Morlet and Mexican-hat, have been used in ECG signal analysis for component detection and disease diagnosis (Stamkopoulos, Diamantaras et al. 1998). Because of the harmonic nature of Morlet and Mexican-hat, they are often used for analysis of harmonic signals. These mother wavelets are not likely to be suitable options in the case of ECG signal classification. In fact, the simplicity of the computed CWT coefficients can be used as a convenient criterion to help in the selection of the mother wavelet as shown below. Figure 7 shows a normal signal and its CWT with different mother wavelets in the scale a=10. Figure 7.a shows a normal signal beat, which has three picks. Figure 7.b shows CWT of the same signal beat with 'Haar' mother wavelet. This figure is very simple and the effects of the raw signal picks are obvious and observable. These effects can be analyzed easily and the extracted features would be suitable and appropriate for the data classification. Also, these computed coefficients can represent morphological differences very well. Figure 7.c shows CWT of the signal with 'Mexican-hat' mother wavelet. The effect of raw signal picks is not obvious in this figure and cannot be analyzed easily. Although this figure is not complicated, the extracted features do not seem to be useful for classification of the data since they are similar to each other. Figure 7.d, 7.e, and 7.f show CWT of the signal with 'Morlet', 'Daubechies8 (db8)' and 'Symlet6 (sym6)' mother wavelets, respectively. It is obvious in these figures that the computed CWT coefficients are similar to each other. Moreover, these figures are quite complicated, and the effects of raw signal picks are not obvious and cannot be analyzed easily. Therefore, the computed CWT coefficients are not suitable features for data classification, since they are similar to each other and cannot represent morphological differences very well. Hence, in this study, 'Haar' mother wavelet has been selected for feature extraction. To compute the CWT of signals, it is not necessary to use scales in the range of 1 through 100. In view of the fact that computing CWT of signals in this range of scales will lead to a huge volume of data as extracted features, it is not advisable to use it. Instead, a specific range of scales, which is suitable and appropriate for feature extraction, is needed. The following is an analysis to determine the appropriate range of scales for the current study. Figure 8 shows 200 samples of a raw normal signal from record 208 from MIT-BIH database and its CWT in different scales, with the 'Haar' mother wavelet. In Figure 8.a, the raw normal signal beat is shown. This signal has 3 picks, which are numbered on the figure; these picks are related to P, R, and T waves. Figure 8.b shows CWT of the signal in scale a=5. In this figure, the noise of the signal has been highlighted; however, the extent of noise is not so large as to interfere with the performance of the neural classifier, and as a result, it is possible to analyze the effect of noise of the raw signal. Moreover, the effect of picks number 1 and 3 can be analyzed to some extent. Figure 8.c shows CWT of the signal in scale a=10. In this figure the effect of the three picks is fully observable and can be analyzed completely; note that there is little noise in the figure. Figure 8.d, which shows CWT of the signal in scale a=20, has no noise and only the effect of three picks can be analyzed according to it. Figures 8.e, 8.f, and 8.g show CWT of signal in scales a=50, 80 and 100, respectively. These figures are similar to each other and neither the noise of the raw signal nor the effect of its picks can be analyzed from these figures; therefore, these figures are not useful for the analysis. It is obvious that morphological differences, which are useful and necessary for neural classifier performance, have been eliminated in these figures. Hence, these extracted features are not appropriate for the neural classifier.

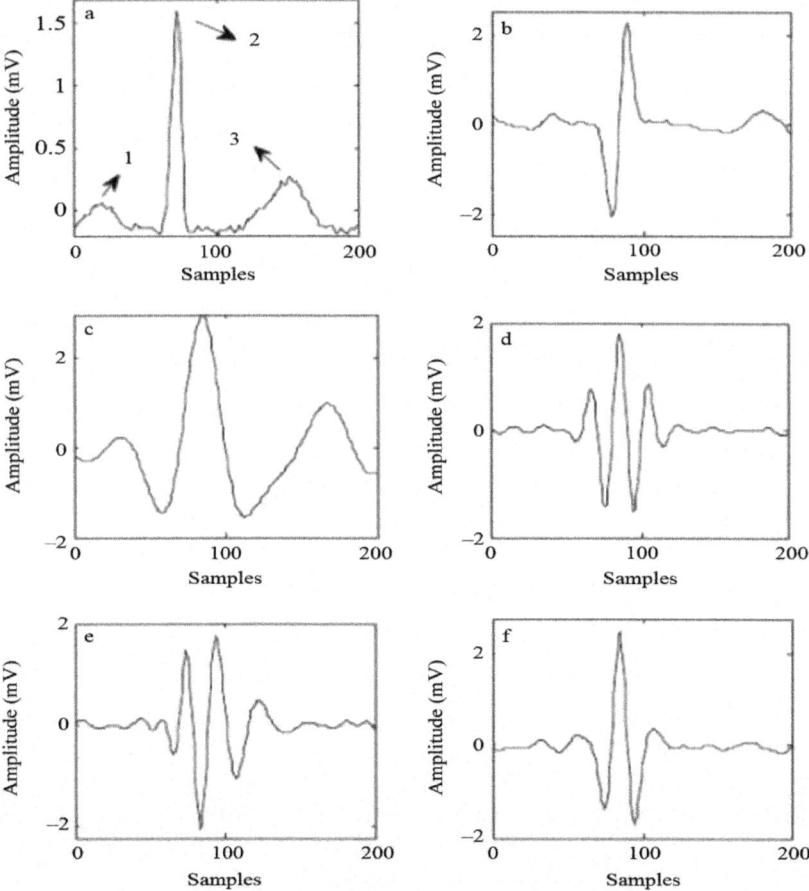

Figure 7: (a) Normal signal beat, (b) CWT of signal with 'Haar' mother wavelet, (c) CWT of signal with 'Mexican hat' mother wavelet, (d) CWT of signal with 'Morlet'

mother wavelet, (e) CWT of signal with 'db8' mother wavelet, (f) CWT of signal with 'sym6' mother wavelet.

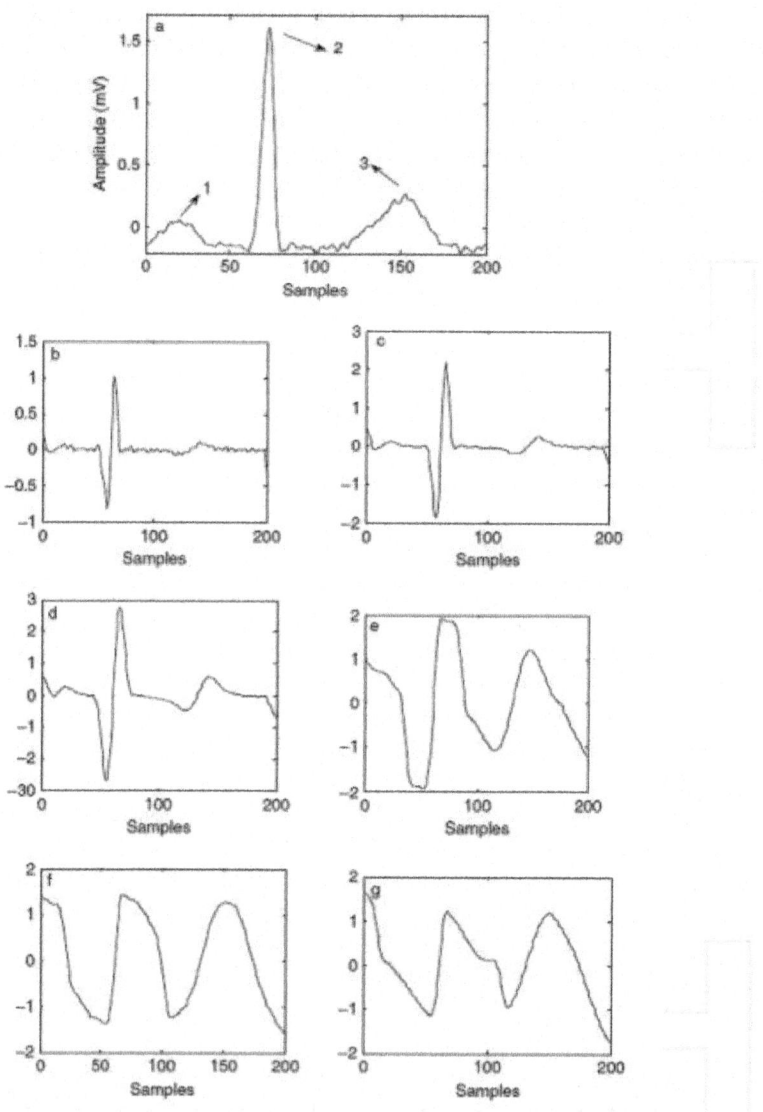

Figure 7: 8. (a) Raw normal signal beat, (b) CWT of signal in scale a=5, (c) CWT of signal in scale a=10, (d) CWT of signal in scale a=20, (e) CWT of signal in scale a=50, (f) CWT of signal in scale a=80, (g) CWT of signal in scale a=100.

From the above analysis, it is clear that computing CWT of the signals in the range of scales from a=5 to 20 can lead to a complete and useful analysis. Since both noise of signals and the effect of morphological differences can be analyzed in this range, the extracted features would be useful for classification of the signals under study

In this study and for the stage of feature extraction, scales in the range of a = 6 through 15 are used that lead to matrices with 10 X 150 dimension for each segment, where each row includes the CWT coefficients in each scale. Using this range of scales has two advantages. First, by computing CWT in the range of a = 6 through 9, the ECG signal can be analyzed in detail. Second, by using the range of a = 10 through 15, the general morphology of the signal and its differences with other types of ECG signals can be highlighted. It should be noted that computing CWT of signals in ten scales can represent morphological differences between several types of ECG signals better than computing CWT of signals in one scale only because of the fact that the differences are analyzed 10 times. This would hence be expected to result in a better performance of the neural classifier. It would not be efficient to use a huge amount of data to perform a pattern recognition process. Hence, in the final level of pre-processing of our algorithm, PCA is applied on the computed matrices of wavelet coefficients, where each of them is a 10x150 matrixes, resulting in 10 principal component (PC) vectors. In this study and for the stage of dimensionality reduction, the first three PC vectors have been selected and arranged as the neural network classifier input vector. This number of PC vectors was chosen according to the results which are presented in Table 1. In this table the accuracy of the neural network classifier with respect to the selected number of PC vectors is shown. According to Table 1, the accuracy of the neural network classifier increases as the number of selected PC vectors increases from 1 to 5, since, by increasing the size of data in this level and this range, the classifier will have a more appropriate set of data for classification. The accuracy of the neural network classifier decreases as the number of selected PC vectors increases from 5 to 10, since at this level, the size of the data is too much for the classifier to have a good performance. Since the difference between classification accuracy in the case of 3 PC vectors and 5 PC vectors is not that significant, we chose 3 PC vectors in order to have a reasonable accuracy, while reducing the computational effort. As a result, by selecting only three PC vectors, dimensionality reduction without significant loss of data information is achieved, leading to a better performance of the neural classifier. These results, which are based on a trial and error method, are not necessarily identical for all kinds of data and all types of algorithm structures. For any change in the algorithm, this analysis should be carried out again in order to find the appropriate number of PC vectors as a classifier input. The prepared vectors, which are the principal components, are used as the neural network classifier input vector. The analysis for providing the input vector structure is the same for both the training and testing database.

Table 1: Variation of classification accuracy with respect to the number of selected PC vectors

Number of Selected PC Vectors	Classification Accuracy (%)
1	98.41 %
2	98.83 %
3	99.17 %
5	99.28 %
8	98.53 %
10	98.94 %

MAIN PROCESS

After finishing the pre-processing stages, data is ready as the input vector for the neural network classifier. In this study, a classical multi-layer perceptron neural network (MLPNN) structure (Silipo and Marchesi 1998; Guler and Ubeyli 2005) is used as the neural network classifier structure. This MLPNN is trained with the back propagation method of error. Selection of the neural network inputs is the most important component of designing the neural network based pattern classification since even the best classifier will perform poorly if the inputs are not selected well (Guler and Ubeyli 2005). The inputs of neural network in this study are constructed in the way which was described in previous section. In our algorithm, we used a classical MLPNN structure with 2 hidden layers and with 60 nodes in the first hidden layer and 15 nodes in the second hidden layer for 160 iterations. The structure of this MLPNN classifier with input, hidden, and output layers is shown in Figure 9. For this structure, the training error was selected to be 0.01 in order to have precise neural network training. From all 6 types of ECG beats under study and for neural network training data, two segments have been selected and processed in the way that was described in previous section

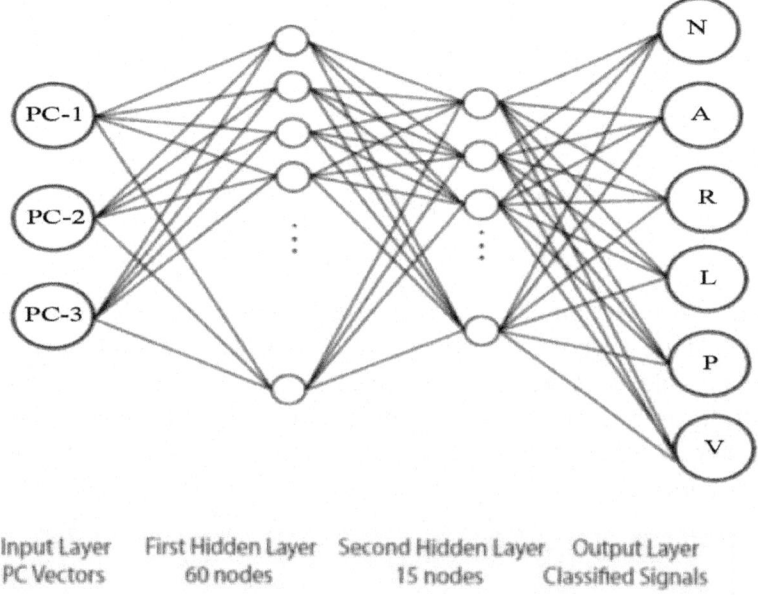

Input Layer   First Hidden Layer   Second Hidden Layer   Output Layer
PC Vectors   60 nodes   15 nodes   Classified Signals

Figure 9: MLPNN structure used as the neural classifier

CLASSIFICATION

When the neural network has been trained, it is ready as a classifier to detect and classify different types of ECG signals into one of six ECG beat groups under study. The classifier has been tested by 100 segments from each group of ECG signals. These testing segments are processed and prepared exactly like the input vector of the neural network; this means that all four levels of pre-processing stage have been applied to each segment in order to prepare it as a testing segment. These segments are used to test and evaluate the trained neural network classifier.

RESULTS

As stated earlier, the MIT-BIH arrhythmia database is used to evaluate the proposed algorithm. To assess the accuracy of the classifier, sensitivity, positive predictive accuracy and total accuracy have been calculated. These are defined as follows:

$$Se = \frac{TP}{(TP + FN)}$$

$$PPA = \frac{TP}{(TP + FP)}$$

$$TA = \frac{TP}{(TP + FN + FP)}$$

Here, TP is the number of true positive detections, FN stands for the number of false negative detections, and FP stands for the number of false positive misdetections. Table 2 shows the result of classification by the neural network. It can be seen from this table that from the whole testing data base, the classification fails only in 5 cases. According to this table, the algorithm achieves a good performance with 99.5 % Se, 99.66% PPA and 99.17% TA.

Table 2: Results of the algorithm on MIT- BIH database

	Normal	Atrial premature beats	Right bundle branch block	Left bundle branch block	Paced	Premature ventricular contraction	Sum
All	100	100	100	100	100	100	600
TP	100	99	99	98	100	99	595
FN	0	0	0	2	0	1	3
FP	0	1	1	0	0	0	2
Se (%)	100	100	100	98	100	99	99.5
PPA (%)	100	99	99	100	100	100	99.66
TA (%)	100	99	99	98	100	99	99.17

A comprehensive comparison between results from different studies in the field of specified ECG beat classification is very difficult since the database, signals under study, the number of arrhythmias in classification, the algorithm structure, and the data processing methods are not the same in the various studies. However, in order to present an estimate of the performance of our algorithm and our classifier we show the results of this study versus the reported results of other well-known studies in the area of selected heart arrhythmias detection in Table 3. As seen from this table, the algorithm in the present study shows reasonably accurate results, and compare favorably with other studies. The goal of this study, which was classification of ECG beats and detection of heart arrhythmias, has clearly been achieved.

Table 3: Comparison of several classifier performances on MIT-BIH database (Blank boxes have not been reported

	TA (%)	PPA (%)	Se (%)
Silipo et al. 1998		85 %	77 %
Papaloukas et al. 2002		89 %	90%
Foo et al. 2002	92 %	-	-
Vikas et al. 2004	-	-	98.02 %
Christov et al. 2004	-	-	99.3 %
Guler et al. 2005	96.94 %	-	96.37 %
Lee et al. 2005	-	-	98.59 %
Kutlu et al. 2008	97.3 %	-	-
Cvikl et al. 2010	-	-	92.36 %
This Study	99.17 %	99.66 %	99.5 %

A NOVEL TECHNIQUE FOR IDENTIFYING PATIENTS WITH ICU NEEDS USING HEMODYNAMIC FEATURES

Modern ICUs are equipped with a large array of alarmed monitors and devices which are used in an attempt to detect clinical changes at the earliest possible moment, so as to prevent any further deterioration in a patient's condition. The effectiveness of these systems depends on the sensitivity and specificity of the alarms, as well as on the responses of the ICU staff to the alarms. However, when large numbers of alarms are either technically false, or true, but clinically irrelevant, response efficiency can be decreased, reducing the quality of patient care and increased patient (and family) anxiety (Laramee, Lesperance et al. 2006). Medical and technical progress has extended the therapeutic possibilities of ICUs tremendously. A multitude of devices is available for monitoring and treatment in an individual assembly according to the requirements of the situation (Friesdorf, Buss et al. 1999). Due to limited physiological monitoring and a patient's individual pathophysiology, intensive care medicine has to cope with a high amount of uncertainty.

Unusual circumstances caused by patients, clinicians and technology occur frequently and must be controlled and managed adequately to prevent a bad outcome and to achieve system reliability (Friesdorf, Buss et al. 1999). Cao et al. (Cao, Eshelman et al. 2008) have used ICU minute-by-minute heart rate (HR) and invasive arterial blood pressure (BP) monitoring trend data collected from the MIMIC II database to predict hemodynamic instability at least two hours before a major clinical intervention. They derived additional physiological parameters of shock index, rate pressure product, heart rate variability, and two measures of trending based on HR and BP and they applied multi-variable logistic regression modeling to carry out classification and implemented validation via bootstrapping, resulting in 75% sensitivity and 80% specificity. Eshelman et al. (Eshelman, Lee et al. 2008) have developed an algorithm for identifying ICU patients who are likely to become hemodynamically unstable. Their algorithm consists of a set of rules that trigger alerts and uses data from multiple sources; it is often able to identify unstable patients earlier and with more accuracy than alerts based on a single threshold.

The rules were generated using the machine learning techniques of support vector machines and neural network, and were tested on retrospective data in the MIMIC II ICU database, yielding a specificity of approximately 90% and a sensitivity of 60%. Several investigations have been reported in the literature in the area of cardiovascular fault diagnosis using hemodynamic features. Javorka et al. (Javorka, Lazarova et al. 2011) compared heart rate and blood pressure variability among young patients with type I diabetes mellitus (DM) and control subjects by using Poincare plots, which are the standard tools of nonlinear dynamic analysis. They found significant reduction of all HRV Poincare plot measure in patients with type I diabetes mellitus, indicating heart rate dysregulation.

The study carried out by Pagani et al. (Pagani, Somers et al. 1988) concerned patients suffering from hypertension. They showed that baroreflex gain decreases with the presence of hypertension. Blasi et al. (Blasi, Jo et al. 2003) studied the effects of arousal from sleep on cardiovascular variability. They performed time-varying spectral analyses of heart rate variability (HRV) and blood pressure variability (BPV) records during acoustically induced arousals from sleep. They found that arousal-induced changes in parasympathetic activity are strongly coupled to respiratory patterns, and that the sympathoexcitatory cardiovascular effects of arousal are relatively long lasting and may accumulate if repetitive arousals occur in close succession. Advances in knowledge-based systems have also enhanced the functionality of intelligent alarm systems and ICU needed patient detection.

Using the knowledge of a domain expert to formulate rules or an expertly classified data set to train an adaptive algorithm has proven useful for intelligent processing of clinical alarms (Laramee, Lesperance et al. 2006). Expert systems such as neural network (Westenskow, Orr et al. 1992), knowledge based decision trees (Muller, Hasman et al. 1997; Tsien, Kohane et al. 2000) and neuro-fuzzy systems (Becker, Thull et al. 1997) that encode the decisions of an expert clinician all show significant statistical improvement in the classification of alarms and ICU needed patients. Singh et al. (Singh and Guttag 2011) proposed a classification algorithm based on a decision tree method for cardiovascular risk stratification. They have shown that the decision tree method can improve performance of the classification algorithm. They have reported that the decision tree models outperform the radial basis function (RBF) kernel-based support vector machine (SVM) classifiers.

Timms et al. (Timms, Gregory et al. 2011) have used a Mock circulation loop for hemodynamic modeling of the cardiovascular system in order to test cardiovascular devices, which are used in the ICU and can provide a better indication of patient's condition for nursing staff. Also, Laramee et al. (Laramee, Lesperance et al. 2006) have described an integrated systems methodology to extract clinically relevant information from physiological data. Such a method would aid significantly in the reduction of false alarms and provide nursing staff with a more reliable indicator of patient condition. Several studies have focused on an effort to find a suitable classifier structure. Ghorbanian et al. (Ghorbanian, Jalali et al. 2011) proposed an algorithm based on a neural network classifier for heart arrhythmias detection.

Their results show that the multi-layer perceptron neural network (MLPNN) structure is a strong and precise classifier. However, they used several pre-processing techniques in their algorithm to improve the performance of the NN classifier. Acharya et al. (Acharya, Bhat et al. 2003) proposed an algorithm based on a neural network classifier and fuzzy cluster for classification of heart arrhythmias. They compared these two classifiers and they reported that the fuzzy cluster is a better classifier in comparison with the neural one. Also, Ozbay et al. (Ozbay, Ceylan et al. 2006) proposed a comparative study of the classification accuracy cardiovascular diseases using a well-known neural network architecture, MLP structure, and a new FCNN for early diagnosis. Based on their test results they suggested that a new proposed FCNN architecture can generalize better than ordinary MLP architecture and also learn better and faster. The method for classification of subjects into two categories of normal and abnormal subjects, as described in this paper, is based on the hypothesis that there should be differences between the hemodynamic data collected from normal subjects and abnormal patients.

This hypothesis is constructed on the same foundation as all developed scoring methods for ICU patients. The idea behind all patient scoring methods in ICU is that critically ill patients in ICU are typically characterized by disturbance of the body's homeostasis. These disturbances can be estimated by measuring to what extent one or many physiologic variables differ from the normal range (Lacroix and Cotting 2005).

METHODOLOGY

While the proposed method in this paper shares some fundamental ideas with traditional scoring methods, it differs from them in two key areas. The first difference comes from fact that the patient scoring methods are based on the wide variety of data ranging from cardiovascular and respiratory systems to neurologic and renal systems variables. However, in our method we use a small subset of hemodynamic data, namely, HR and systolic blood pressure (SBP). The principal objection to this could be that such a small amount of data could be insufficient for identifying the patient state; the answer to this objection leads us to the second major difference of the proposed method with the scoring methods. Scoring methods just look at the data as they are being collected in the ICU, and ignore information hidden in the different time scales. In our proposed method on the other hand, this hidden information is extracted which can be expected to give us better insight into the patient's physiological condition. The data used in this study is collected from the Physionet database. Data are collected from two databases: MIT-BIH Polysmonographic and MIMIC II databases within Physionet archive. Twenty five subjects from these databases were collected for training. For each subject, ECG signal and blood pressure waveform, in a five-hour range of the total data were collected. For the first part of the study, the HR and SBP series for each subject are derived from ECG and arterial pressure waveforms respectively. The algorithm of the developed method of this study is shown in Figure (10). According to the proposed algorithm, in the first step and after collecting the data, four features which highlight the differences between normal subjects and patients, are extracted from data. We then define four criteria based on the extracted features. These four criteria which form the basis of our classification algorithm are: circle criterion, estimation error criterion, Poincare care plot deviation, and autonomic response delay criterion. In the next step and for the task of classification, we define three groups; namely, healthy, high risk and patient. Then we design three fuzzy membership functions for each criterion to find the subject degree of membership to each group. Finally, a scoring method is developed based on the degree of membership of each case, and subjects are classified based on this scoring method.

In the following sections, we provide a step by step description of our method, beginning with the definition of the proposed criteria.

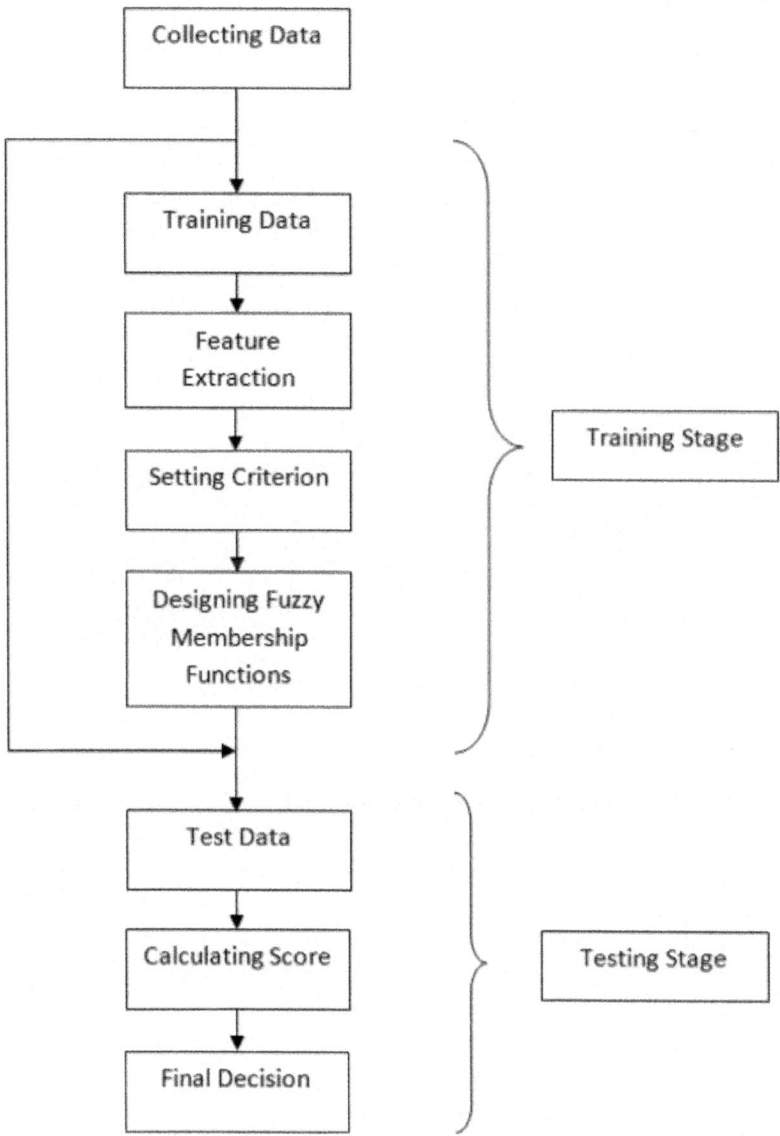

Figure 10: Schematic of the proposed algorithm. The proposed algorithm consists of two stages: training and testing. In the training stage 25 subjects' data are used to extract features to classify patients from healthy subjects. In the test stage subjects will be divided into three predefined groups of healthy, high risk and patient, based on their assigned score.

Circle criterion

To evaluate the differences between healthy and patients, the SBP against HR diagram for each subject is plotted. Figure 11 shows these plots for healthy and patient cases. Clearly, the plots show a significant difference between normal subjects and abnormal patients: the data for normal subjects are concentrated, while those of the patients are scattered. The mean value of SBP and HR for each normal subject and abnormal patient is then calculated and plotted in one diagram. Figure 12 shows the mean values for all the subjects in one diagram. The principal difference between the two groups is quite clear. This

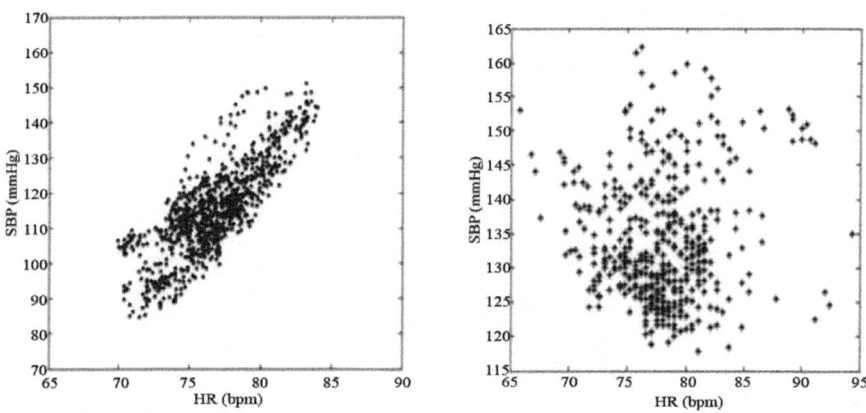

Figure 11: SBP against HR for a healthy (left) and an abnormal (right) case

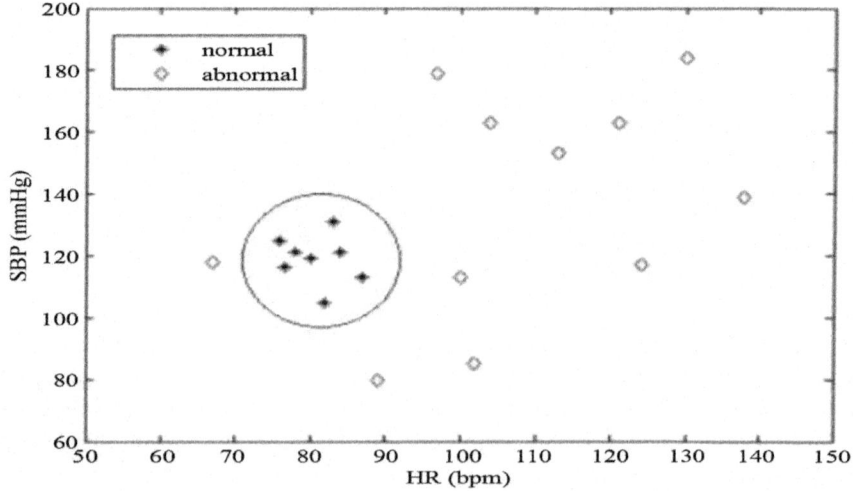

Figure 12: Mean values of SBP versus HR for all subjects

diagram reveals the fact that there are differences between the HR and SBP data in normal subjects and abnormal ones. The plot shows that the data for the normal subjects is clustered and limited in a specific area, while those of the patients are spread out through the whole plot. The first criterion is named the "circle criterion". The center of the circle is located at point "O" where its coordinates are the mean values of HR and SBP of normal patients and, in this case, is (83, 120). The radius of this circle is calculated based on Euclidian distance between the center and the outer limit of the circle. A given subject would be considered to be a patient if its corresponding means (HR, SBP) point is out of the healthy subject's circle (the limited area).

Estimation error criterion

As the second feature, a system identification method is used for the prediction of the next HR based on the current and previous HR and SBP data. A Nonlinear ARX or NARX model is employed to estimate HR series (Jalali, Ghaffari et al. 2011). NARX models in general are represented by the following equation:

$$y(t) = F\big(y(t-1), y(t-2), ..., y(t-n_a), u(t-n_k), ..., u(t-n_k-n_b+1)\big)$$

where, y(t) and u(t) are the output and input of the system, respectively. In Eq. (1) the matrix $[n_a \quad n_b \quad n_k]$ is the same as the order of the model. Model order is selected by use of the A-Information Criterion (AIC) method. This is the traditional method for model order selection in cardiovascular system identification research. Model order for data in this research has been calculated to be [9,6,3].

In this criterion, Artificial Neuro Fuzzy Inference System (ANFIS) structure is employed for the identification. The model has 15 inputs and one output. Membership functions for inputs are designed based on physiological facts. Since the nervous system consists of sympathetic and parasympathetic nerves, for each input, two generalized bell-shaped membership functions are assigned to designate the sympathetic and parasympathetic functions. The system identification results are described in Table 4. The results in this table show that differences exist in the normalized root mean square error (NRMSE) with respect to the estimation of the HR for the two groups under study. In particular, the results indicate that NRMSE is smaller for normal subjects than for patients. These differences are due to the fact that the model is designed for normal subjects; thus, the output of the model for patients have higher errors than for normal subjects.

Table 4: Error estimation for identification of HR baroreflex

Group	Mean	Max	Min
Normal	0.193	0.238	0.119
abnormal	0.367	0.473	0.263

Based on these results and noting that the maximum error for healthy subject is 0.238, while the minimum error for patient is 0.263, we define a second criterion called "estimation error criterion". According to this criterion, the subject would be flagged as abnormal if the calculated error in HR estimation raise is more than 0.25.

Poincare Plot Deviation

A Poincare plot, named after Henri Poincare, is used to quantify self-similarity in processes which are usually characterized by periodic functions. This plot is commonly used in heart rate variability (HRV) analysis. The Poincare plot is a graph in which each heart rate episode is plotted as a function of previous HR, and then the line y=x is fitted to the data. In (Javorka, Lazarova et al. 2011) this method is also applied to classify patients with type I DM from healthy subjects. Drawing the Poincare plot for healthy and abnormal subjects, it is found that the deviation from the mentioned line in healthy subjects is less than in abnormal subjects. These plots are shown in Figure 13. The deviation from the line y=x in the Poincare plot for the two groups under study is shown in Table 5. Therefore, we define the third criterion using this deviation to characterize abnormality. Based on this criterion, subjects would be called abnormal If deviation from line y=x is more than 15%.

Table 5: Deviation from line y=x in Poincare plot

Group	Mean	Max
Healthy	8%	13%
Patient	19%	24%

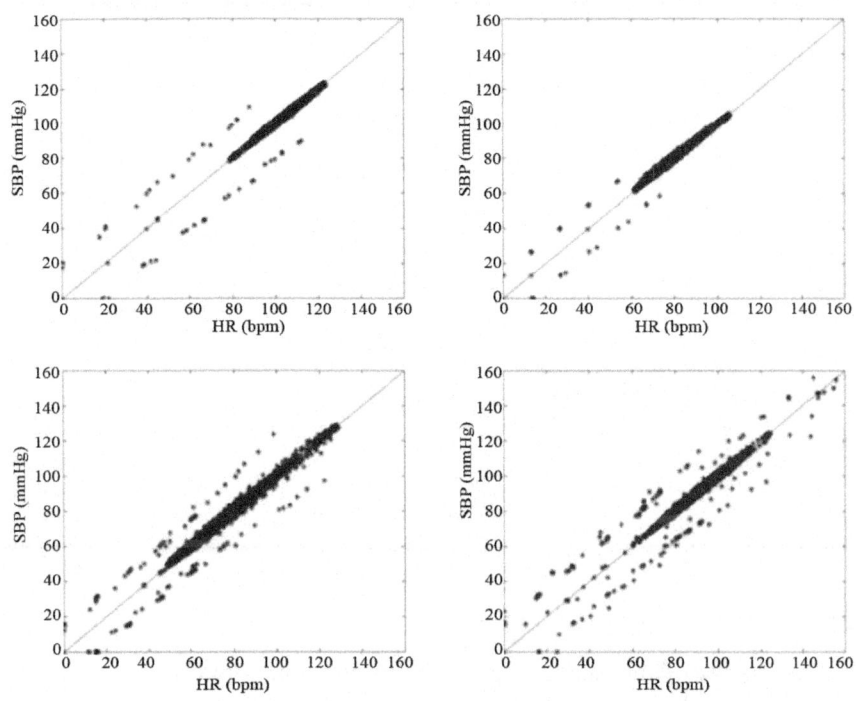

Figure 13: Poincare plots of HR for two healthy (up) and two abnormal (down) cases. The Poincare plot is a plot of HR(n+1) vs. HR(n). Line y=x is illustrated in all pictures.

Autonomic response delay criterion

The normally occurring delay in the autonomic response to a stimulus has its origins in the parasympathetic nervous system. Calculating the delay for healthy subjects and patients we can infer that response delays in abnormal subjects are remarkably higher than healthy subjects. The results of calculating the delay in the autonomic response are shown in Figure 14. Fifteen abnormal patients and ten healthy subjects were involved in the training group.

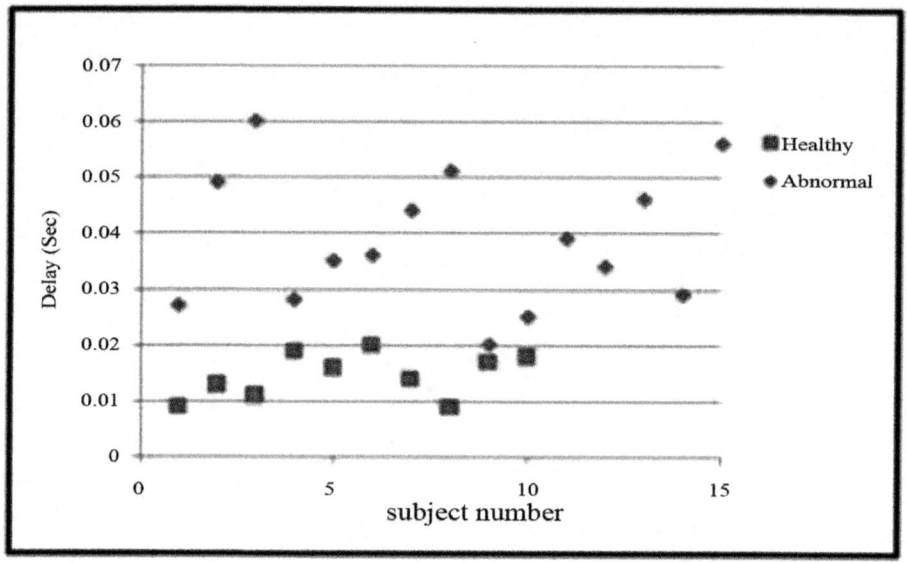

Figure 14: Delay in autonomic response

The results of the delay calculations in the autonomic response are also represented in Table 6. Based on the above results, we define the fourth criterion where the subject is characterized as abnormal if the calculated delay in the autonomic response increases to more than 0.021 second.

Table 6: delay in autonomic response for two groups

Group	Mean Delay (sec)	Max Delay (sec)
Healthy	0.015	0.02
Patient	0.038	0.06

After deriving the four criteria discussed above, an algorithm is designed to classify healthy subjects from patients. In the following section we describe the proposed algorithm.

Scoring method and classification algorithm

Based on the evaluated criteria from training data, an algorithm is developed to automatically distinguish patients from healthy subjects. The algorithm is based on a fuzzy decision making method. First, for each criterion, three Gaussian bell membership functions are designed as an indicator of three major groups: healthy, high risk and patient. Since this algorithm is designed for clinical use and since there exists a high degree of uncertainty in clinical applications, we added the high risk groups to our predefined healthy and patient groups to account the cases that do not completely belong to the healthy or patient groups. For the training part we first made a general guess for the shape of the membership functions. The membership functions during the training round then adapt their shape parameters to the incoming data for best classification performance. Now the classifier is designed and ready for the testing stage. Figure 15 represents the adapted membership functions for each criterion based on the training data. To test the developed algorithm, in the first step for each subject, all the mentioned features that form the basis of four criteria are extracted and used as an input for the four abnormality criteria. Then, for each criterion, the subject's degree of membership to all groups is evaluated. In this step, for each subject, we have 12 degrees of membership to the designed three groups, meaning four degrees of membership for each group. After evaluating the degree of memberships, the cumulative sum of the four degrees of membership of each group will be calculated. In this stage we have three numbers indicating subject's degree of membership to each group. We call these numbers the subject's "score" for each group. A given subject will belong to the group whose score is the largest

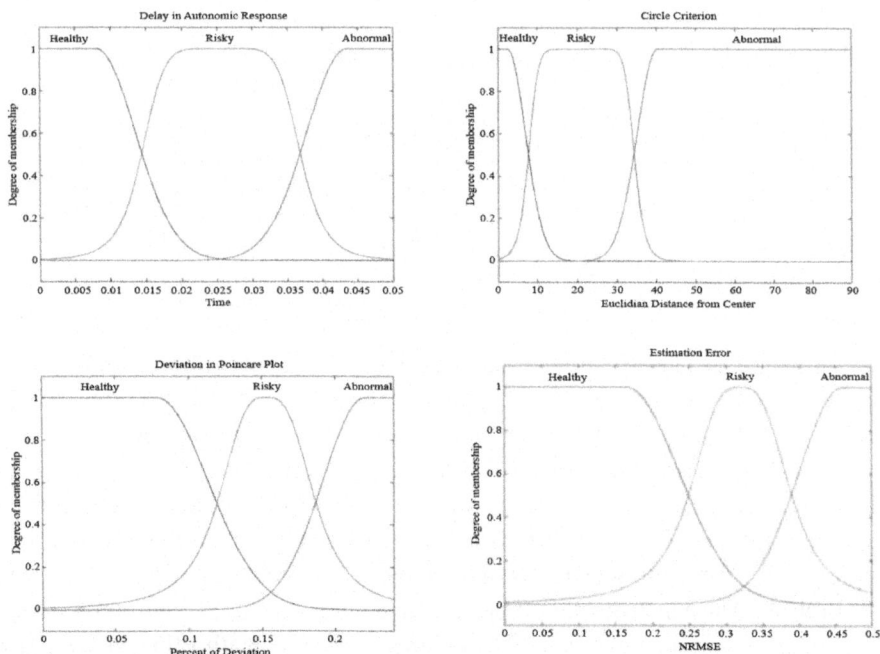

Figure 15: The designed membership functions for each criterion

Results

From a total of seventy subject data which were collected from MIMIC II database, the algorithm was first trained with twenty five subjects including ten healthy and fifteen patients. The training data was selected randomly to avoid bias toward a specific disease. Then, three groups of subjects were tested, each group with four healthy individuals and eleven patients.

The proposed method was applied to 45 cases from Physionet database, containing 12 healthy subjects and 33 patients. From all cases, 37 cases were accurately detected, while there was one false detection. Furthermore, in five cases, a patient subject was classified as high risk and, in two cases, a healthy subject was classified as high risk. Here, TP is the number of true positive detections, FN stands for the number of false negative detections, and FP stands for the number of false positive misdetections. Table (7) shows the overall result of the classification for all 45 cases of the 3 groups. The FP is the healthy subject who is misclassified as a high risk subject and FN is the patient who is misclassified as a high risk subject. According to this table, the scoring method of the proposed algorithm results in 86% sensitivity, 94.8% positive predictive accuracy and 82.2% total accuracy.

Table 7: Results of testing the algorithm on Physionet database

Group	I	II	III	All
All	15	15	15	45
TP	12	13	12	37
FN	3	1	2	6
FP	0	1	1	2
Se (%)	80	92.8	85.7	86
PPA (%)	100	92.8	92..3	94.8
TA (%)	80	86	80	82.2

A comprehensive comparison between the results of different studies in the field of identifying ICU needed patients by the use of hemodynamic features is very difficult since the database, signals under study, the algorithm structure, and the data processing methods are not the same in the various studies. However, in order to present an estimate of the performance of our algorithm and our classifier we show the results of this study versus the reported results of two other well-known studies in the area of ICU needed patients identifying in Table 8. As seen from this table, the algorithm in the present study shows reasonably accurate results, and compares favorably with other studies. The goal of this study, which was identifying patients with ICU needs by use of the hemodynamic features, has clearly been achieved.

Table 8: Comparison of several classifier performances on MIMIC II ICU database (Blank boxes have not been reported)

Study	Se (%)
Cao et al. [1]	75
Eshelman et al. [4]	60
This study	86

CONCLUSION

Physiological time series, including hemodynamic and electrophysiological data clearly represent the physiological state of subjects in a medical environment. Automatic detection of heart arrhythmias could be very important in clinical usage and lead to early detection of a fairly common malady and could help contribute to reduced mortality as cardiovascular disease remains the leading cause of death around the world.

Hemodynamic instability is most commonly associated with abnormal or unstable blood pressure (BP), especially hypotension, or is more broadly associated with inadequate global or regional perfusion. Inadequate perfusion may compromise important organs, such as heart and brain, due to limits on coronary and cerebral autoregulation and cause life-threatening illnesses or even death. Therefore, it is crucial to identify patients who are likely to become hemodynamically unstable for an early detection and treatment of these life-threatening conditions. In the first example of this study, the use of neural networks for classification of the ECG beats is presented. Several stages of pre-processing have been used in order to prepare the most appropriate input vector for the neural classifier.

ECG signal baseline wandering is one of the most critical problems for neural classifiers, since it causes virtual morphological differences between same types of ECG beats. In this example, this wandering is removed by application of a signal filtering method which leads to better results. As the performance of the computerized ECG classification algorithms depends on the selection of the ECG features, continuous wavelet transform, which performs better than other methods, is used to extract appropriate features. Also, data dimensionality reduction is one of the most important ways of improving neural classification, since large volume of data causes problems for neural network classifier performance, and reduction in the data size is necessary for better performance of the classifier. Therefore, principal component analysis is used to achieve dimensionality reduction. Results show that PCA is more effective than other reported methods. The performance of the proposed algorithm has been shown to be reasonably acceptable and ECG beat detection and classification has been achieved. Compared to other reported work in this field, the presented algorithm shows reasonably accurate results in the field of heart arrhythmia detection. The main advantage of this example is that, by using ten scales in computing CWT of signals, the morphological differences between several types of ECG signal are highlighted and the extracted features show the differences more clearly. Another advantage of this example is that the reduction of the dimension of data by applying PCA led to the most appropriate input vector for neural network classifier which improved the performance of the neural network classifier significantly. The main achievement of this algorithm is that the classifier in this example detects 6 types of ECG signals which include normal beats and 5 types of arrhythmia beats. Even though the number of ECG signal types considered in this example is much larger than the typical number of ECG signal types in other studies in this field, the classification results lead to a reasonably good performance. In the second example of this study, a scoring method based on fuzzy logic and feature extraction is proposed to distinguish patients from healthy subjects. The method is based on the same principle that the ICU scoring methods follow: that of finding differences between hemodynamic data of healthy subjects and patients. Four different criteria are proposed to detect and identify patients from a group of subjects. For each criterion a fuzzy classifier is designed such that the individuals are classified into the healthy, high risk and patient fuzzy groups. In other words, a given person may have a membership grade in all three classes. A score is assigned to the subject for that group which is defined as the sum of degree of memberships to one group for different criteria. The algorithm calculates a combined criterion based on the results of the four criteria to arrive at a classification decision for each individual. It is shown that the algorithm is highly reliable and has been able to detect correctly all members of the first group. It is also been able to detect all eleven patients in each of the next two groups correctly. Only one of the healthy members in the second and third was classified as high risk. In this example, four different criteria were proposed and used in the proposed algorithm in order to detect the abnormalities in testing subjects. From each testing subject, various features were extracted and used as input for the criteria, and based on the results of all four criteria, a decision was made about the type of subject, as to whether he/she is normal, high risk or a patient. The proposed algorithm gave reliable results in detecting the ICU needed patients but still needs to be improved. The difference between the proposed method in this example and other similar research in this field of study is that by using the presented algorithm in this example, existence of any abnormality in a patient will be found, while in most similar studies in this area, a specific abnormality is found in a patient or among a database of subjects. Therefore, our results are more general and more useful from the point of view of clinical applications. This method tends to be more detective rather than predictive, and this could be one drawback of the algorithm. Further investigations need to be carried out to render the algorithm more predictive

REFERENCES

1.   Acharya, U. R., P. S. Bhat, S. S. Iyengar, A. Rao and S. Dua (2003). Classification of heart rate data using artificial neural network and fuzzy equivalence relation. Pattern Recognition Vol.36, No.1, (Jan 2003), pp. 61-68

2.   Addison, P. S. (2005). Wavelet transforms and the ECG: a review. Physiological Measurement Vol.26, No.5, (Oct 2005), pp. R155-R199

3.   Addison, P. S., J. N. Watson, G. R. Clegg, M. Holzer, F. Sterz and C. E. Robertson (2000). Evaluating arrhythmias in ECG signals using wavelet transforms. IEEE Eng Med Biol Mag Vol.19, No.5, (Sep-Oct 2000), pp. 104-109

4.   Becker, K., B. Thull, H. KasmacherLeidinger, J. Stemmer, G. Rau, G. Kalff and H. J. Zimmermann (1997). Design and validation of an intelligent patient monitoring and alarm system based on a fuzzy logic process model. Artificial Intelligence in Medicine Vol.11, No.1, (Sep 1997), pp. 33-53

5.   Blasi, A., J. Jo, E. Valladares, B. J. Morgan, J. B. Skatrud and M. C. Khoo (2003). Cardiovascular variability after arousal from sleep: time-varying spectral analysis. J Appl Physiol Vol.95, No.4, (Oct 2003), pp. 1394-1404

6.   Cao, H., L. Eshelman, N. Chbat, L. Nielsen, B. Gross and M. Saeed (2008). Predicting ICU hemodynamic instability using continuous multiparameter trends, Conf Proc IEEE Eng Med Biol Soc, pp. 3803-3806, Vancouver, Canada, August 21-23, 2008

7.   Caudill, M. (1989). Neural Networks Primer, Miller Freeman Publications, San Francisco, USA

8.   Ceylan, R. and Y. Ozbay (2007). Comparison of FCM, PCA and WT techniques for classification ECG arrhythmias using artificial neural network. Expert Systems with Applications Vol.33, No.2, (Aug 2007), pp. 286-295

9.   Christov, I. and G. Bortolan (2004). Ranking of pattern recognition parameters for premature ventricular contractions classification by neural networks. Physiological Measurement Vol.25, No.5, (Oct 2004), pp. 1281-1290

10. Cvikl, M. and A. Zemva (2010). FPGA-oriented HW/SW implementation of ECG beat detection and classification algorithm. Digital Signal Processing Vol.20, No.1, (Jan 2010), pp. 238-248

11. Daubechies, I. (2006). Ten Lectures on Wavelet, SIAM, Philadelphia

12. Dokur, Z. and T. Olmez (2001). ECG beat classification by a novel hybrid neural network. Comput Methods Programs Biomed Vol.66, No.2-3, (Sep 2001), pp. 167-181 Eshelman, L. J., K. P. Lee, J. J. Frassica, W. Zong, L. Nielsen and M. Saeed (2008).

13. Development and evaluation of predictive alerts for hemodynamic instability in ICU patients. AMIA Annu Symp Proc2008), pp. 379-383

14. Foo, S. Y., G. Stuart, B. Harvey and A. Meyer-Baese (2002). Neural network-based ECG pattern recognition. Engineering Applications of Artificial Intelligence Vol.15, 2002), pp. 253-260

15. Friesdorf, W., B. Buss and M. Gobel (1999). Monitoring alarms--the key to patient's safety in the ICU? Intensive Care Med Vol.25, No.12, (Dec 1999), pp. 1350-1352

16. Ghaffari, A., H. SadAbadi and M. Ghasemi (2006). A Mathematical algorithm for ECG Signals Denoising Using Window Analysis. Biomed Papers Vol.151, No.73-78, 2006)

17. Ghorbanian, P., A. Jalali, A. Ghaffari and C. Nataraj (2011). An improved procedure for detection of heart arrhythmias with novel pre-processing techniques. Expert Systems2011),

18. Guler, I. and E. D. Ubeyli (2005). Adaptive neuro-fuzzy inference system for classification of EEG signals using wavelet coefficients. J Neurosci Methods Vol.148, No.2, (Oct 30 2005), pp. 113-121

19. Guler, I. and E. D. Ubeyli (2005). ECG beat classifier designed by combined neural network model. Pattern Recognition Vol.38, No.2, (Feb 2005), pp. 199-208

20. Jalali, A., A. Ghaffari, P. Ghorbanian and C. Nataraj (2011). Identification of sympathetic and parasympathetic nerves function in cardiovascular regulation using ANFIS approximation. Artif Intell Med Vol.52, No.1, (May 2011), pp. 27-32

21. Javorka, M., Z. Lazarova, I. Tonhajzerova, Z. Turianikova, N. Honzikova, B. Fiser, K. Javorka and M. Baumert (2011). Baroreflex analysis in diabetes mellitus: linear and nonlinear approaches. Med Biol Eng Comput Vol.49, No.3, (Mar 2011), pp. 279-288

22. Jolliffe, I. T. (2002). Principal Component Analysis, Springer, New York Kundu, M., M. Nasipuri and D. K. Basu (2000). Knowledge-based ECG

interpretation: a critical review. Pattern Recognition Vol.33, No.3, (Mar 2000), pp. 351-373 Kutlu, Y., D. Kuntalp and M. Kuntalp (2008).

23. Arrhythmia classification using higher order statistics. IEEE Signal Processing, Communication and Applications Conference. Turkey: 1-4. Lacroix, J. and J. Cotting (2005).

24. Severity of illness and organ dysfunction scoring in children. Pediatr Crit Care Med Vol.6, No.3 Suppl, (May 2005), pp. S126-134 Laramee, C. B., L. Lesperance, D. Gause and K. McLeod (2006). Intelligent alarm processing into clinical knowledge. Conf Proc IEEE Eng Med Biol Soc Vol.Suppl, 2006), pp. 6657- 6659 Lee, J., K. Park, M. Song and K. Lee (2005).

25. Arrhythmia classification with reduced features by linear discriminant analysis. Conf Proc IEEE Eng Med Biol Soc Vol.2, 2005), pp. 1142-1144

26. Maglaveras, N., T. Stamkopoulos, K. Diamantaras, C. Pappas and M. Strintzis (1998). ECG pattern recognition and classification using non-linear transformations and neural networks: a review. Int J Med Inform Vol.52, No.1-3, (Oct-Dec 1998), pp. 191-208

27. Muller, B., A. Hasman and J. A. Blom (1997). Evaluation of automatically learned intelligent alarm systems. Computer Methods and Programs in Biomedicine Vol.54, No.3, (Nov 1997), pp. 209-226

28. Osowski, S. and T. H. Linh (2001). ECG beat recognition using fuzzy hybrid neural network. IEEE Trans Biomed Eng Vol.48, No.11, (Nov 2001), pp. 1265-1271

29. Owis, M. I., A. H. Abou-Zied, A. B. Youssef and Y. M. Kadah (2002). Study of features based on nonlinear dynamical modeling in ECG arrhythmia detection and classification. IEEE Trans Biomed Eng Vol.49, No.7, (Jul 2002), pp. 733-736 Ozbay, B. and B. Karlýk (2001).

30. A recognition of ECG arrhythmias using artificial neural network, Annual Conference of IEEE EMBS, pp. 1680-1683, Istanbul, Turkey, 2001 Ozbay, Y., R. Ceylan and B. Karlik (2006).

31. A fuzzy clustering neural network architecture for classification of ECG arrhythmias. Computers in Biology and Medicine Vol.36, No.4, (Apr 2006), pp. 376-388 Pagani, M., V. Somers, R. Furlan, S. Dell'Orto, J. Conway, G. Baselli, S. Cerutti, P. Sleight and A. Malliani (1988).

32. Changes in autonomic regulation induced by physical training in mild hypertension. Hypertension Vol.12, No.6, (Dec 1988), pp. 600-610 Papaloukas, C., D. I. Fotiadis, A. Likas and L. K. Michalis (2002).

33. An ischemia detection method based on artificial neural networks. Artif Intell Med Vol.24, No.2, (Feb 2002), pp. 167-178 Saxena, S. C., V. Kumar and S. T. Hamde (2002).

34. Feature extraction from ECG signals using wavelet transforms for disease diagnostics. International Journal of Systems Science Vol.33, No.13, (Oct 20 2002), pp. 1073-1085

35. Silipo, R. and C. Marchesi (1998). Artificial neural networks for automatic ECG analysis. Ieee Transactions on Signal Processing Vol.46, No.5, (May 1998), pp. 1417-1425

36. Singh, A. and J. V. Guttag (2011). A Comparison of Non-symmetric Entropy-based Classification trees and Support Vector Machine for Cardiovascular Risk Stratification, Annual Conference of the IEEE EMBS, pp. 79-82, Boston, MA USA, 2011

37. Stamkopoulos, T., K. Diamantaras, N. Maglaveras and M. Strintzis (1998). ECG analysis using nonlinear PCA neural networks for ischemia detection. Ieee Transactions on Signal Processing Vol.46, No.11, (Nov 1998), pp. 3058-3067

38. Timms, D. L., S. D. Gregory, M. C. Stevens and J. F. Fraser (2011). Hemodynamic Modeling of Cardiovascular System Using Mock Circulation Loops to Test Cardiovascular Devices, Annual Conference of the IEEE EMBS, pp. 4301-4304, Boston, MA USA, 2011

39. Tsien, C. L., I. S. Kohane and N. McIntosh (2000). Multiple signal integration by decision tree induction to detect artifacts in the neonatal intensive care unit.

40. Artificial Intelligence in Medicine Vol.19, No.3, (Jul 2000), pp. 189-202 Vikas, J. and J. S. Sahambi (2004). Neural network and wavelets in arrhythmia classification. Asian Applied Computing Vol.32, 2004), pp. 92-99 Wang, X. C. and K. K. Paliwal (2003).

41. Feature extraction and dimensionality reduction algorithms and their applications in vowel recognition. Pattern Recognition Vol.36, No.10, (Oct 2003), pp. 2429-2439

42. Westenskow, D. R., J. A. Orr, F. H. Simon, H. J. Bender and H. Frankenberger (1992). Intelligent alarms reduce anesthesiologist's response time to critical faults. Anesthesiology Vol.77, No.6, (Dec 1992), pp. 1074-1079 Zadeh, L. A. (1968).

43. Fuzzy Algorithms. Information and Control Vol.12, No.2, 1968), pp. 94-102 Zhang, H. and L. Q. Zhang (2005). ECG analysis based on PCA and Support Vector Machines, IEEE International Conference on Neural Networks and Brain, pp. 743-747, Beijing, China, 2005

Chapter 2

THE SUCCESSIVE ZOOMING GENETIC ALGORITHM AND ITS APPLICATIONS

Young-Doo Kwon[1] and Dae-Suep Lee[2]

[1]School of Mechanical Engineering & IEDT, Kyungpook National University,
[2]Division of Mechanical Engineering, Yeungjin College, Daegu, Republic of Korea

INTRODUCTION

Optimization techniques range widely from the early gradient techniques 1 to the latest random techniques 16, 18, 19 including ant colony optimization 13, 17. Gradient techniques are very powerful when applied to smooth well-behaved objective functions, and especially, when applied to a monotonic function with a single optimum. They encounter certain difficulties in problems with multi optima and in those having a sharp gradient, such as a problem with constraint or jump. The solution may converge to a local optimum, or not converge to any optimum but diverge near a jump. To remedy these difficulties, several different techniques based on random searching have been developed: full random methods, simulated annealing methods, and genetic algorithms. The full random methods like the Monte Calro method are perfectly global but exhibit very slow convergence. The simulated annealing methods are modified versions of the hill-climbing technique; they have enhanced global search ability but they too have slow convergence rates. Genetic algorithms 2-5 have good global search ability with relatively fast convergence rate. The global search ability is relevant to the crossover and mutations of chromosomes of the reproduced pool. Fast convergence is relevant to the selection that takes into account the fitness by the roulette or tournament operation. Micro-GA 3 does not need to adopt mutation, for it introduces completely new individuals in the mating pool that have no relation to the evolved similar individuals. The pool size is smaller than that used by the simple GA , which needs a big pool to generate a variety of individuals. Versatile genetic algorithms have some difficulty in identifying the optimal solution that is correct up to several significant digits. They can quickly approach to the vicinity of the global optimum, but thereafter, march too slowly to it in many cases. To enhance the convergence rate, hybrid methods have been developed. A typical one obtains a rough optimum using the GA first, and then approaches the exact optimum by using a gradient method. Other one finds the rough optimum using the GA first, and then searches for the exact optimum by using the GA again in a local domain selected based on certain logic 7. The SZGA (Successive Zooming Genetic Algorithm) 6, 8-12 zooms the search domain for a specified number of steps to obtain the optimal solution. The tentative optimum solutions

are corrected up to several significant digits according to the number of zooms and the zooming rate. The SZGA can predict the possibility that the solution found is the exact optimum solution. The zooming factor, number of sub-iteration populations, number of zooms, and dimensions of a given problem affect the possibility and accuracy of the solution. In this chapter, we examine these parameters and propose a method for selecting the optimal values of parameters in SZGA.

The Successive Zooming Genetic Algorithm

This section briefly introduces the successive zooming genetic algorithm 6 and provides the basis for the selection of the parameters used. The algorithm has been applied successively to many optimization problems. The successive zooming genetic algorithm involves the successive reduction of the search space around the candidate optimum point. Although this method can also be applied to a general Genetic Algorithm (GA), in the current study it is applied to the Micro-Genetic Algorithm (MGA).

The working procedure of the SZGA is as follows. First, the initial solution population is generated and the MGA is applied. Thereafter, for every 100 generations, the elitist point with the best fitness is identified. Next, the search domain is reduced to $(X_{OPT}-\alpha^k/2, \ X_{OPT}+\alpha^k/2)$, and then the optimization procedure is continued on the reduced domain (Fig. 1). This reduction of the search domain increases the resolution of the solution, and the procedure is repeated until a satisfactory solution is identified.

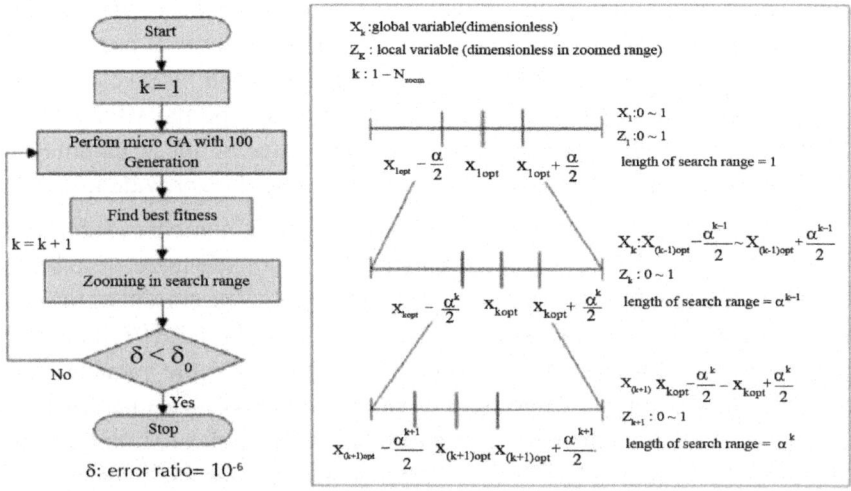

Figure 1: Flowchart of SZGA and schematics of successive zooming algorithm The SZGA can assess the reliability of the obtained optimal solution by the reliability equation expressed with three parameters and the dimension of the solution N_{VAR}.

$$R_{SZGA} = [1 - (1 - (\alpha/2)^{N_{VAR}} \times \beta_{AVG})^{N_{SP}}]^{N_{ZOOM}-1}$$

(1)

Where

α: zooming factor,   β: improvement factor
N_{VAR}: dimension of the solution, N_{ZOOM}: number of zooms
N_{SUB}: number of sub-iterations, N_{POP}: number of populations
N_{SP}: total number of individuals during the sub-iterations ($N_{SP}=N_{SUB}\times N_{POP}$)

Three parameters control the performance of the SZGA: the zooming factor $^\lrcorner$, number of zooming operations N_{ZOOM}, and sub-iteration population number N_{SP}. According to previous research, the optimal parameters for S_{ZGA}, such as the zooming factor, number of zooming operations, and sub-iteration population number, are closely related to the number of variables used in the optimization problem.

Selection of parameters in the SZGA The zooming factor α, number of sub-iteration population N_{SP}, and number of zooms NZOOM of SZGA greatly affect the possibility of finding an optimal solution and the accuracy of the found solution. These parameters have been selected empirically or by the trial and error method. The values assigned to these parameters determine the reliability and accuracy of the solution. Improper values of parameters might result in the loss of the global optimum, or may necessitate a further search because of the low accuracy of the optimum solution found based on these improper values. We shall optimize the SZGA itself by investigating the relation among these parameters and by finding the optimal values of these parameters. A standard way of selecting the values of these parameters in SZGA, considering the dimension of the solution, will be provided. .

The SZGA is optimized using the zooming factor α, number of sub-iteration population N_{SP}, and the number of zooms N_{ZOOM}, for the target reliability of 99.9999% and target accuracy of 10^{-6}. The objective of the current optimization is to minimize the computation load while meeting the target reliability and target accuracy. Instead of using empirical values for the parameters, we suggest a standard way of finding the optimal values of these parameters for the objective function, by using any optimization technique, to find the optimal values of these parameters which optimize the SZGA itself. Thus, before trying to solve any given optimization problem using SZGA, we shall optimize the SZGA itself first to find the optimal values of its parameters, and then solve the original optimization problem to find the optimal solution by using these parameters. After analyzing the relation among the parameters, we shall formulate the problem for the optimization of SZGA itself. The solution vector is comprised of the zooming factor α, the number of sub-iteration population N_{SP}, and the number of zooms N_{ZOOM}. The objective function is composed of the difference of the actual reliability to the target reliability, difference of the actual accuracy to the target accuracy, difference of the actual N_{SP} to the proposed N_{SP}, and the number of total population generated as well.

$$F(\alpha,\ N_{SP},\ N_{ZOOM}) = \Delta R_{SZGA} + \Delta A + \Delta N_{SP} + (N_{SP} \times N_{ZOOM})$$

(2)

where,

ΔRS_{ZGA} : difference to the target reliability

ΔA : difference to the target accuracy

ΔN_{SP} : difference to the proposed N_{SP} The problem for optimzation of SZGA itself can be formulated by using this objective function as follows:

Minimize F(X)   (3)

Where

$$X = \{\alpha, \, N_{SP}, \, N_{ZOOM}\}^T$$

$$0 < \alpha < 1$$
$$N_{SP} \sim 100$$
$$N_{ZOOM} > 1$$

The difference of the actual reliability to the target reliability is the difference between R_{SZGA} and 99.9999%, where reliability R_{SZGA} is rewritten with an average improvement factor as

$$R_{SZGA} = [1 - (1 - (\alpha / 2)^{N_{VAR}} \times \beta_{AVG})^{N_{SP}}]^{N_{ZOOM} - 1}$$

(4)

Here, we can see the average improvement factor β_{AVG}, which is to be regressed later on. The difference of realized accuracy to the target accuracy is the difference between accuracy A and 10^{-6}, where accuracy A is actually the upper limit and may be written as,

$$A = \alpha^{N_{ZOOM} - 1}$$

(5)

The difference of the actual NSP to the proposed NSP is difference between NSP and 100 [7]. In organizing the optimization algorithm, each element in the objective function is given different weights according to its importance. Thus, the target reliability and target accuracy are met first, and then the number of total population generated is minimized. Although any optimization technique could have been used to slove eq.(3), one can adopt the SZGA in optimizing the SZGA itself to obtain a solution fast and accurately. The parameters in SZGA have been optimized by using the objective function and improvement factor averaged after regression for a test function [9]. The target reliability is 99.9999% and target accuracy of solution is 10^{-6}. The proposed number of sub-iteration population N_{SP} is 100. Table 1 shows the optimized values for the SZGA parameters for four cases of different number of design variables. We found a similar tendency to Table 1 for test functions of various numbers of design variables. We also found that the recommended number of sub-iteration population NSP would no longer be acceptable to assure reliability and accuracy for the cases whose number of design variables is over 1. A much greater number of sub-iteration population is needed to obtain an optimal solution with the proper reliability (99.9999%) and accuracy (10^{-6}). To confirm our optimized result, we fixed two parameters in the feasible domain that satisfy the target reliability and target accuracy, and checked the change in the objective function as a function of the remaining parameter. Examples of the change in the objective function for the case of four design variables showed the validity of the obtained optimal values of the

Table 1. Result of optimized parameters in SZGA for different number of design variables

No. of Variables	2	4	8	16
Zooming Factor α	.02573	.1303	.4216	.5176
N_{ZOOM}	5	8	17	22
N_{SP}	1,000	2,000	9,510	1,479,230
No. of Function Evaluation	5,000	16,000	161,670	32,543,060

parameters. Although these values may not be valid for all the other cases, they can be used as a good reference for new problems. Some other ways of choosing the values of these parameters will be given later on.

Programming for successive zooming and pre-zoning algorithms

Programming the SZGA is simple, as explained below. This zooming philosophy may not be confined only in GA, but can be applied to most other global search algorithms. Let $Y(I)$ be the global variables ranging $YMIN(I)$ ~ $YMAX(I)$, where I is the design variable number. $Z(I)$ consists of local normalized variables ranging 0~1. Thus, the relation between them is as follows in FORTRAN;

```
      DO 10 I=1,NVAR ! NVAR=NO. of VARIABLES
10 Y(I)=YMIN(I)+(YMAX(I)-YMIN(I))*Z(I)
```

The relation between local variable $Z(I)$ and local variable $X(I)$ (0~1) in the zoomed region is as follows;

```
      DO 12 I=1,NVAR
12 Z(I)=ZOPT(I,JWIN)+ALP**(JWIN-1)*(X(I)-0.5)
```

Where, ZOPT(I,JWIN) is the elitist in the zoom step (JWIN-1), and ALP is the zooming factor. Note that ZOPT(I,JWIN-1) is more logical. However, the argument is increased by one to meet old versions of FORTRAN, which require a positive integer as a dimension argument. Based on the elitist in step (JWIN-1), we are seeking variables in step JWIN. Please note that ZOPT(I,1)=0. A pre-zoning algorithm adjusts the gussed initial zone to a very reasonable zone after one set of generation.

```
DO 14 I=1,NVAR
   YMIN(I)=YINP(I)-BTA*ABS(YINP(I))
14 YMAX(I)=YINP(I)+BTA*ABS(YINP(I))
```

Where, YINP(I)is the elitist obtained after one set of generation. Thus, we eliminate the assumed initial boundary, and establish a new reasonable boundary. The coefficient BTA may be properly selected, say 0.5.

HYBRID GENETIC ALGORITHM

Genetic algorithms are stochastic global search methods based on the mechanism of natural selection and natural reproduction. GAs have been applied to structural optimization problems because they can solve optimization problems that involve mixing continuous, discontinuous, and non-convex regions etc. The SGA (simple GA) has been improved to MGA by using some techniques like tournament selection as well as the elitist strategy. Yet, GAs have some difficulty in fast searching the exact optimum point at a later stage. The DPE (Dynamic Parameter Encoding) GA [4] uses a digital zooming technique, which does not change a digit of a higher rank further after a certain stage. The SZGA (Successive Zooming GA) zooms the searching area successively, and thus the convergence rate is greatly increased. A new hybrid GA technique, which guarantees to find the optimum point, has been proposed [7, 14.]

The hybrid GA first identifies a quasi optimal point using an MGA, which has better searching ability than the simple genetic algorithm. To solve the convergence problem at the later stage, we employed hybrid algorithms that combine the global GA with local search algorithms (DFP 1 or MGA). The hybrid algorithm using the DFP(Davidon Fletcher Powell) method incorporates the advantages of both a genetic algorithm and the gradient search technique. The other hybrid algorithm of global GA and local GA at the zoomed area is called LGA (Locally zoomed GA), checks the concavity condition near the quasi minimum point. The enhancement of the above hybrid algorithms is verified by application of these algorithms to the gate optimization problem.

In this hybrid algorithm of minimization problem, an MGA is performed generation-bygeneration until there is no further change of the objective function, and then the approximate optimum solution is found at Z_{MCA}. The gradients of the objective function as a function of the design variables are checked, if the concavity condition 1 is satisfied at the boundary of a small zoomed area (Fig. 2). If the condition is not satisfied, the small zoomed area is increased by δ. After several iterations, concavity conditions are finally achieved at the boundary of the final zoomed area ($\kappa\delta \times \kappa\delta$) centered at ZMCA. With the elitist solution from the global GA (approximate optimum solution, Z_{MCA}) and the concavity condition, the optimum point is found within the final zoomed area [$Z(i) : (Z_{MCA}(i) - \kappa\delta) \sim (Z_{MCA}(i) + \kappa\delta)$]. From this point, a local GA is performed for the small finally zoomed area, which probably contains the optimum point. Usually, this area is much smaller than the original are, so the convergence rate increases considerably (note that the first approximate solution prematurely converged to an inexact but near optimum point).

Water gates need to be installed in dams to regulate the flow-rate and to ensure the containing function of dams. Among these gates, the radial gate is widely used to regulate the flow-rate of huge dams because of its accuracy, easy opening and closing, endurance etc. Moreover, 3-arm type radial gate has better performance than 2-arm type, in connection with the section size of girders and the vibration characteristics during discharging operation. Table 2 compares the optimized results for a 3-arm type radial gate, which considers the reactions to the minimized main weight of the structure including vertical girders with or without arms. The hybrid algorithm (MGA+DFP, MGA+LGA) obtained the exact optimal solution of 0.690488E+10 after far fewer generations of 4100 than the 9000 by MGA, which result in a close but not the exact solution of 0.690497E+10.

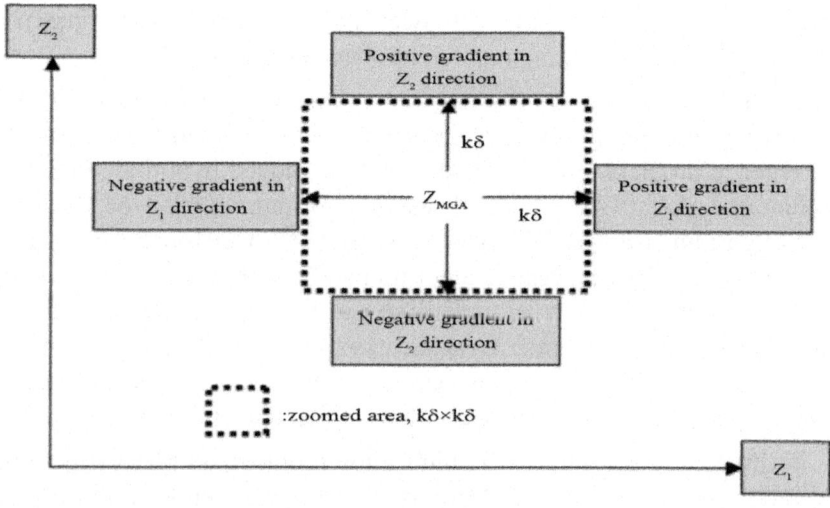

Fig. 2. Confirmed zoomed region after checking the concavity condition

Table 2. Comparison of results: MGA, MGA+DFP, MGA+LGA

3-arm type	Micro GA	MGA+DFP	MGA+LGA
Convergence Generation	9000	4000+α	4100
Objection Function	0.690497E+10	0.690488E+10	0.690488E+10

Example of the SZGA The value of the zooming factor α, an optimal parameter was obtained in reference [8], and was found to show good match with the empirical one. Using this zooming factor in SZGA, the displacement of a truss structure was derived by minimizing the total potential energy of the system. The capacity of the servomotor, which operates the wicket gate mounted in a Kaplan type turbine of the electric power generator, was optimized using SZGA with the value of zooming factor [8]. This is just one parameter among the full optimal parameters discussed in sec.2.1 9. Therefore, the analysis done with this factor [8] is a simplified analysis. As commented in section 2.1, the values of the parameters of a well-behaved test model suggested in the Table 1 can be used for an optimization, or the values of the parameters obtained in another way as discussed in the next section can be used. Several additional examples of SZGA optimization are presented in the following sections to provide more insight on SZGA and to find another way of choosing the values of the SZGA parameters. The first example finds the Moony-Rivlin coefficients of a rubber material to compare with those from the least square method. The second example is a damage detection problem in which the difference between the measured natural frequencies and those of the assumed damage in the structure is minimized. The third example finds the optimal link specification (lengths and initial angular positions of members) to control the double link system with one motor in an automotive diesel engine. The fourth and last example finds an optimal specification (parametric sizes at specified positions) of a ceramic jar that satisfies the required holding capacity.

Determination of Mooney-Rivlin coefficients

The rubber is a very important mechanical material in everyday life, used widely in mechanical engineering and automotive engineering. Rubber has low production cost and many advantages such as its characteristic softness, processability, and hyper-elasticity. The development of the rubber parts including most process of the shape design, product process, test evaluation, ingredient blending for the required property has used the empirical methods. CAE based on advances in computer-aided structural analysis software is applied to many products. FEM method is applied on various models of rubber parts to evaluate the non-linearity property and the theoretical hyper-elastic behavior of rubber, and to develop analysis codes for large, non-linear deformation. The structure of rubber-like materials are difficult to analyze because of their material nonlinearity and geometric non-linearity as well as their incompressibility. Furthermore, unlike other linear materials, rubber materials have hyper-elasticity, which is expressed by the strain energy function. The representative strain energy functions in the finite element analysis of rubber are the extension ratio invariant function (Mooney-Rivlin model) and the principal extension ratio function (Ogden model). This case uses the Mooney-Rivlin model to investigate the behavior of a rubber material. The value of the zooming factor changes according to the number of variables and the population number of a generation. If the population number is large, more exact solution can be obtained than the approach with smaller one. For a large population number, which is inevitable in the case of many design variables, longer computation time is needed. In this case, because six design valuables are used to solve the six material properties, nine hundred population units per one generation are used. At this time, whenever zooming is needed, the function is calculated 90,000 times, where, 900 is the population number per one generation and 100 is generation number per one zooming because zooming is implemented after 100 generations . So the point number searched per one valuable is 6 units ($=90,000^{1/6}$). To search the optimum point, the zooming factor must be not less than 1/6. Therefore, the zooming factor of 0.2 is used.

The maximum generation number must be decided after the zooming factor is chosen. If the zooming factor is large, the exact solution can be solved as increasing zooming step. Generation numbers have to be decided by the user because they affect the amount of calculation like the population numbers do. For example, when zooming factor of 0.3 is chosen and Maxgen (maximum allowed generation number) is decided as 1000 (N_{ZOOM} = 10), the accuracy of the final searching range becomes $Z_{RANGE} = (N^{zoom-1}) = 0.3(10-1) = 1.97E-05$, and if Maxgen is decided by 1500 (NZOOM = 15) the final searching range becomes $Z_{RANGE} = (N^{zoom-1}) = 0.3(15-1) = 4.78E-08$, where Z_{RANGE} is the value related with the resolution of solution and is the searching range after N steps of zooming. The smaller this value is, the more exact the solution becomes. In this case, Maxgen=900 is adopted. S_{ZGA} minimized the total error better than the other two methods

Table 3. Comparisons of errors among the different methods for obtaining Mooney-Rivlin 6 coefficients

Errors to be minimized	Haines & Wilson	Least Square	SZGA
Simple extension	0.757932	0.709209	0.921277
Pure shear	0.702015	0.620089	0.370579
Equi-biaxial	13.2580	0.242475	0.139983
Total error	14.7180	1.57177	1.43184

Damage Detection Of Structures

Structures can sometimes experience failures far earlier than expected, due to fabrication errors, material imperfections, fatigue, or design mistakes, of which fatigue failure is perhaps the most common . Therefore, to protect a structure from any catastrophic failure, regular inspections that include knocking, visual searches, and other nondestructive testing are conducted. However, these methods are all localized and depend strongly on the skill and experience of the inspector. Consequently, smart and global ways of searching for damages have recently been investigated by using rational algorithms, powerful computers, and FEM. The objective function of the difference between the measured data and the computed data is minimized according to an assumed structural damage to find the locations and intensities of possible damages in a structure. The measured data can be the displacement of certain points or the natural frequencies of the structure, while the computed data are obtained by FEM using an assumed structural damage, whose severity is graded between 0 and 1. For example, Chou et al. used static displacements at a few locations in a discrete structure composed of truss members, and adopted a kind of mixed string scheme as an implicit redundant representation. Meanwhile, Rao adopted a residual force method, where the fitness is the inverse of an objective function, which is the vector sum of the residual forces, and Koh adopted a stacked mode shape correlation that could locate multiple damages without incorporating sensitivity information 11. Yet, a typical structure can be sub-divided into many finite elements and has many degrees of freedom. Thus, FEM for a static analysis, as well as for a frequency analysis, takes a long time. For a GA, the analysis time is related to the number of functions used for evaluating fitness. This number can become uncontrollable when monitoring a full structure, and as a result, the RAM or memory space required becomes too large and the access rate too slow when handling so much data. Accordingly, the proposed SZGA is very effective in this case, as it does not require so many chromosomes, even as few as 4, thereby overcoming the slow-down of the convergence rate of the conventional GA, which need many chromosomes in determining the extent of a damage. Furthermore, the issue of many degrees of freedom can also be solved by subdividing the monitoring problem into smaller sub-problems because the number of damages will likely be between 1~4, as long as the structure was designed properly. Moreover, the fact that cracks usually initiate at the outer and tensile stressed locations of a structure is also an advantage. As a result, the number of sub-problems becomes manageable, and the required time is much reasonable. Several tests were performed first to determine the effectiveness of the SZGA for structure monitoring, where regional zooming is not necessary. Next, the procedure used to subdivide the monitoring problem is presented, along with a comparison of the amount of computation required between a full-scale monitoring analysis and a sub-- divide monitoring analysis according to the number of probable damage sites. The optimization problem for various cases of structural damage detection was solved by using three or six variables, zooming factor of 0.2 or 0.3, and total number of function evaluations of 100,000 or 150,000, which is $N_{ZOOM} \times$ sub-iteration population number. The sub-iteration population number means the total population number in a sub-generation of one zooming.

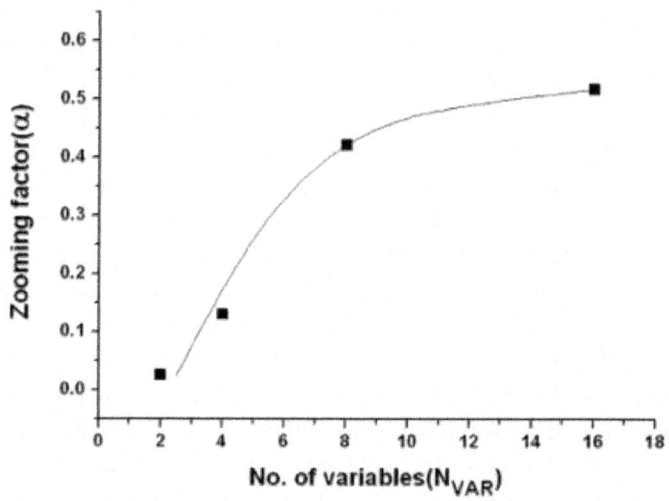

Fig. 3. Zooming factor with respect to the number of variables

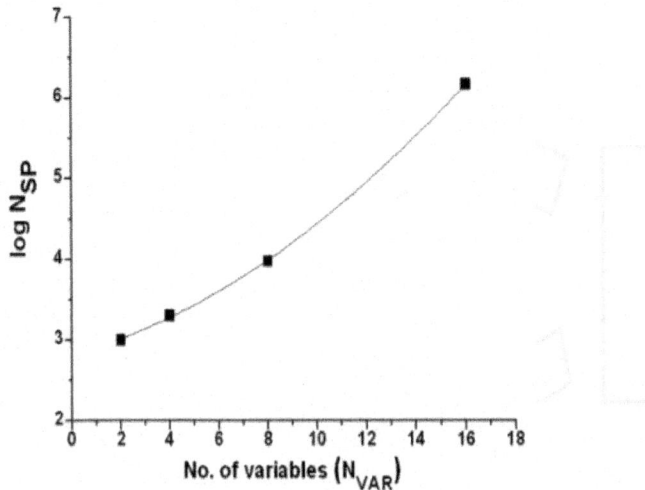

Fig. 4. Number of sub-iteration population with respect to the number of variables

Fig. 3, Fig. 4 and Fig. 5 are the fitting curves of '
'$N_{VAR} - \alpha$', '$N_{VAR} - N_{SP}$' and 'N_{VAR} - Number of function calculation' relationship data, respectively, based on Table 1. These figures are prepared for the data point not shown in Table 1 for interpolation purpose.

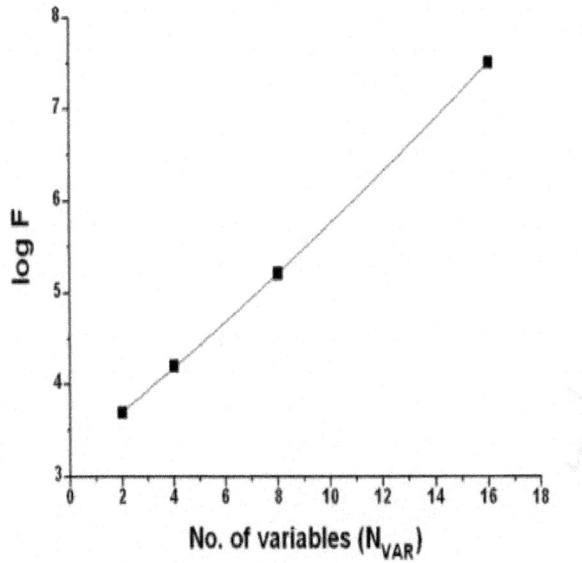

Fig. 5. Number of function calculations with respect to the number of variables The SZGA can pinpoint an optimal solution by searching a successively zoomed domain.

Yet, in addition to its fine-tuning capability, the SZGA only requires several chromosomes for each zoomed domain, which is a very useful characteristic for structural damage detection of a large structure that has a great number of solution variables. In the present study, just four or six digits of chromosomes were used. The accuracy of optimal solution is guaranteed by the successively zoomed infinitesimal range. Most structures have few cracks, which may exist at different locations. Therefore, a combinational search method is suggested to search for separate cracks by choosing probable damage site as $_nC_k$. n denotes the number of total elements and k denotes the number of possible crack sites (1~4). Thus, up to four cracks (k) were considered in a continuum structure modelled with n (= 20) elements, and the number of function calculations between the combinational search and the full scale search was compared.

$$_nC_k = \frac{n!}{k!(n-k)!}$$

(6)

Table 4. Result of combinational searching method to reduce amount of calculation in SZGA

No. of cracks	$_nC_k$	No. of function calculation		Ratio (Combinational/Full)
		Combinational search	Full scale search	
1	20	0.580671×10^5	0.578096×10^9	0.100445×10^{-3}
2	190	0.950000×10^6	0.578096×10^9	0.164332×10^{-2}
3	1140	0.990843×10^7	0.578096×10^9	0.171398×10^{-1}
4	4845	0.740788×10^8	0.578096×10^9	0.128143

When monitoring the entire structure, the number of function calculations became about six hundred million based on the relation between the number of variables and the number of

function calculations. However, when the combinational searching method was used, the number of function calculations was reduced by about 10^{-1}~10^{-4} times when compared to the full-scale monitoring case, as shown in Table 4. Table 5 shows the good detection of the damage using the combination method and SZGA.

Table 5. Result of structural damage detection using the combination method and SZGA

Element No.	19	20	25	26	31	32
Actual soundness factor	1	1	0.5	1	1	1
Damage detection result	1.0	1.0	0.499999	1.0	1.0	1.0

LINK SYSTEM DESIGN USING WEIGHTING FACTORS

This section presents a procedure involving the use of a genetic algorithm for the optimal designs of single four-bar link systems and a double four-bar link system used in diesel engines. Studies concerning the optimal design of the double link system comprised of both an open single link system and a closed single link system which are rare, and moreover the application of the SZGA in this field is hard to find, where the shape of objective function have a broad, flat distribution 12. During the optimal design of single four-bar link systems, one can find that for the case of equal IO angles, the initial and final configurations show certain symmetry. In the case of open single link systems, the radii of the IO links are the same and there is planar symmetry. In the case of closed single systems, the radii of the IO links are the same and there is point symmetry. To control the Swirl Control Valve in small High Speed Direct Injection engines, there are two types of actuating systems. The first uses a single DC motor controlled by Pulse Width Modulation, while the second uses two DC motors. However, this study uses the first type of actuator for the simultaneous control of two Swirl Control Valves using a double link system. When two intake valves in a diesel engine are controlled by a single motor, they usually exhibit quite different angular responses when the design variables for the control link system are not properly selected. Therefore, in order to ensure balanced performance in diesel engines with two intake valves, an optimization problem needs to be formulated and solved to find the best set of design variables for the double four-bar link system, which in turn can be used to minimize the different responses to a single input. Two weighting factors are introduced into the objective function to maintain balance between the multi-objective functions. The proper ratios of weighting factors between objective functions are chosen graphically. The optimal solutions provided by the SZGA and developed FORTRAN Link programs can be confirmed by monitoring the fitness. The reduction in the objective functions is listed in the tables. The responses of the output links that follow the simultaneously acting input links are verified by experiment and the Recurdyn 3-D kinematic analysis package. The experimental and analysis results show good correspondence.

The proposed optimal design process was successfully applied to a recently launched luxury Sports Utility Vehicle model. Table 6 shows the original response and that of the optimized model. The optimal model exhibits almost the exact left and right outputs, and the difference between the left and right responses of 0.603 is thought to be a least value for the given positions of the link centers and the double control system adopting a single input motor.

Table 6. Comparison of original and optimal models

Model	Input (degree)	Output(degree)		
		Left	Right	Max. Difference
Original	0-90	0-89.144	0-91.958	2.044
Optimal	0-90	0-89.999	0-89.999	0.603

Proper band width for equality constraints

In a problem having an equality constraint, it is not so simple for GA to satisfy the constraint while maintaining efficiency. Optimal solution lies on the line of equality constraint. It is very important to gernerate individuals on or near the equality line. However, the desirable narrow area including the equality line is very small compared with the whole area. The number of individual generated in this narrow area is much less than those in the outer area of the desirable narrow area including the equality line. Therefore, the convergence rate of GA or SZGA is significantly slow for the problems with equality constraints. The bandwidth method is proposed to overcome this kind of slow convergence rate. For the minimization problems, we added a basic penalty function to meet the equality constraint, which will be explained soon. For this problem with the basic constraint, we can not expect a rapid convergence rate as mentioned above. Therefore, we added an additional penalty function to the region, located out of the desirable narrow area including the equality line, to make an infeasible area of a very highly increased objective function. The bandwidth denotes the half width of the narrow region with the basic penalty only. There are three methods to handle the equality constraints using GA.

One is to give both sides the penalty functions along the equality condition. The other is to give one side the monotonic function and other side the even (jump) penalty function along the equality constraint. However, the one side with the monotonic penalty should be feasible. And, the final one is to apply one side with no penalty function and the other side with the even (jump) penalty function along the equality constraint, and the one side of no penalty function should be feasible. The penalty methods provided in Fig. 6 only with original penalty, is the basic technique for handling the equality constraint [15]. With this kind of basic technique only, however, the convergence rate would be too slow to reach the optimal point. Many generated individuals are wasted because they mostly too far from the equality constraint line. Therefore we need an additional penalty function to increase the effectiveness of GA. That is an additional penalty to the objective function if the condition is located in outer region of a certain bandwidth centered with the equality constraint.

Using the type (c) equality constraint and additional bandwidth penalty, the design of a ceramic jar was optimized for three values of zooming factors and various bandwidths of equality constraint, as shown in Fig. 7 and Table 7. The result showed a proper range of bandwidth for the equality constraint. In Table 7, the optimal solutions were found for the jar, satisfying the equality constraint of 2 liter volume.

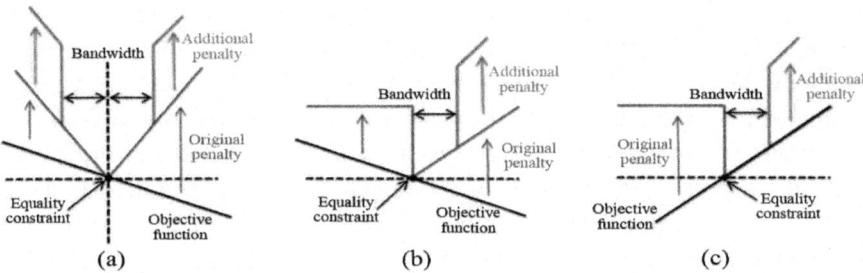

Fig. 6. Three methods to handle the equality constraint in GA.

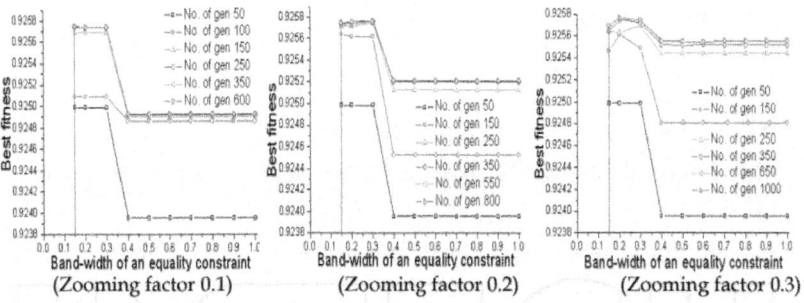

Fig. 7. Best fitness for band-width of an equality constraint and numbers of generation.

Table 7. Proper bandwidths and the optimal solutions for three zooming factors

Zooming factors	Proper band-width	Weight (kg)	Volume (liter)	Z_1	Z_2
0.1	0.15~0.3	0.0802	2.000	0.4790	1.000
0.2	0.15~0.3	0.0802	2.000	0.4790	1.000
0.3	0.15~0.3	0.0802	2.000	0.4790	1.000

This optimization problem does not converge below 0.15 of the band-width of an equality constraint, because the objective function is rather complicated and the band-width is relatively too narrow to give the most candidated optimal individual out of feasible region.

When the band-width is bigger than about 0.3, the best fitness dropped rapidly. In other words, if we open the full range as the feasible solution range, the optimal ridge would be too narrow to be chosen by GA. In conclusion, a too narrow bandwidth may lead to a divergence and a too wide bandwidth may result in inefficiency.

FURTHER STUDIES AND CONCLUDING REMARKS

The SZGA explained in the foregoing sections may be applied to more fields of interest, such as, the optimal design of ceramic pieces considering important factors like beauty, usage, stability, strength, lid, and exact volume. Prediction of a long -term performance of a rubber seal installed in an automotive engine is another possible application. The most dominant characteristics of SZGA are its accuracy up to the required significant digits, and its rapid convergence rate even in the later stage. However, users have to properly select the parameters, namely, the zooming factor, number of zooms, and number of sub-domain population. A useful reference can be found in Table 1, Fig. 3, Fig. 4, and Fig. 5. The number of zooms can be determined by eq.(5) for a given upper limit of accuracy. The number of sub-domain population has been recommended as a fixed number until now, however, it may be varied as a function of the zooming step.

REFERENCES

1.   D.M. Himmelblau, 1972, Applied Nonlinear Programming, McGraw-Hill.
2.   D.E. Goldberg, 1989, Genetic Algorithms in Search, Optimization, and Machine Learning,Addison-Wesley.

3. K. Krishnakumar, 1989, "Micro-genetic algorithms for stationary and non-stationary.function optimization," SPIEP, Intelligent Control and Adaptive Systems, Vol. 1196,.pp. 289~296.

4. N.N. Schraudolph, R.K. Belew, 1992, "Dynamic parameter encoding for genetic.algorithms," Journal of Machine Learning, Vol. 9, pp. 9-21.

5. D.L. Carroll, 1996, "Genetic algorithms and optimizing chemical oxygen-Iodine lasers,".Developments in Theoretical and Applied Mechanics, Vol. 18, pp. 411~424.

6. Y.D. Kwon, S.B. Kwon, S.B. Jin and J.Y. Kim, 2003, "Convergence enhanced genetic.algorithm with successive zooming method for solving continuous optimization.problems," Computers and Structures, Vol. 81, Iss. 17, pp. 1715~1725.

7. Y.D. Kwon, S.B. Jin, J.Y. Kim, and I.H. Lee, 2004, "Local zooming genetic algorithm and.its application to radial gate support problems," Structural Engineering and.Mechanics, An International Journal, Vol. 17, No. 5, pp. 611~626.

8. Y.D. Kwon, H.W. Kwon, J.Y. Kim, S.B. Jin, 2004, "Optimization and verification of.parameters used in successive zooming genetic algorithm," Journal of Ocean.Engineering and Technology, Vol. 18, No. 5, pp. 29~35.

9. Y.D. Kwon, H.W. Kwon, S.W. Cho, and S.H. Kang, 2006, "Convergence rate of the.successive zooming genetic algorithm using optimal control parameters," WSEAS.Transactions on Computers, Vol. 5, Iss. 6, pp. 1200~12007.

10. Y.D. Kwon, J.Y. Kim, Y.C. Jung, and I.S. Han, 2007, "Estimation of rubber material.property by successive zooming genetic algorithm," JSME, Journal of Solid.Mechanics and Materials Engineering, Vol. 1, Iss. 6, pp. 815-826.

11. Y.D. Kwon, H.W. Kwon, W.J. Kim, and S.D. Yeo, 2008, "Structural damage detection in.continuum structures using successive zooming genetic algorithm," Structural.Engineering and Mechanics, An International Journal, Vol. 30, No. 2, pp. 135~146.

12. Y.D. Kwon, C.H. Sohn, S.B. Kwon, and J.G. Lim, 2009, "Optimal design of link systems using successive zooming genetic algorithm," SPIE, Progress in Biomedical Optics and Imaging, Vol. 7493, No. 1~3, pp. 17-1~8.

13. O. Baskan, S. Haldenbilen, Huseyin Ceylan, Halim Ceylan, 2009, "A new solution.algorithm for improving performance of ant colony optimization," Applied.Mathematics and Computation, Vol. 211, Iss. 1, pp. 75~84..[14] N. Tutkun, 2009, "Optimization of multimodal continuous

functions using a new.crossover for the real-coded genetic algorithms," Expert Systems with Applications,.Vol. 3, Iss. 4, pp. 8172~8177.

14.   Y.D. Kwon, S.W. Han, and J.W. Do, 2010, "Convergence rate of the successive zooming.genetic algorithm for band-widths of equality constraint," International Journal of.Modern Physics B, Vol. 24, No. 15&16, pp. 2731~2736.

15.   Z. Ye, Z. Lee, M. Xie, 2010, "Some improvement on adaptive genetic algorithms for.reliability-related applications," Reliability Engineering and System Safety, Vol. 95,.Iss. 2, pp. 120~126.

16.   K. Wei, H. Tuo, Z. Jing, 2010, "Improving binary ant colony optimization by adaptive.pheromone and commutative solution update," IEEE, 5th International Conference.on Bio Inspired Computing: Theory and Applications (BIC-TA), pp. 565~569.

17.   S. Babaie-Kafaki, R. Ghanbari, N. Mahdavi-Amiri, 2011, "Two effective hybrid.metaheuristic algorithms for minimization of multimodal functions," Computer.Mathematics, Vol. 88, Iss. 11, pp. 2415~2428.

18.   M.A. Ahandani, N.P. Shirjoposh, R. Banimahd, 2011, "Three modified version of.differential evolution for continuous optimization," Soft Computing, Vol. 15, Iss. 4,.pp. 803~830

Chapter 3

PUBLIC PORTFOLIO SELECTION COMBINING GENETIC ALGORITHMS AND MATHEMATICAL DECISION ANALYSIS

Eduardo Fernández-González, Inés Vega-López and Jorge Navarro-Castillo

Autonomous University of Sinaloa México

INTRODUCTION

A central and frequently contentious issue in public policy analysis is the allocation of funds to competing projects. Public resources for financing social projects are particularly scarce. Very often, the cumulative budget being requested ostensibly overwhelms what can be granted. Moreover, strategic, political and ideological criteria pervade the administrative decisions on such assignments (Peterson, 2005). To satisfy these normative criteria, that underlie either prevalent public policies or governmental ideology, it is obviously convenient both to prioritize projects and to construct project-portfolios according to rational principles (e.g., maximizing social benefits). Fernandez et al. (2009a) assert that public projects may be characterized as follows.

- They may be undoubtedly profitable, but their benefits are indirect, perhaps only longterm visible, and hard to quantify.

- Aside from their potential economic contributions to social welfare, there are intangible benefits that should be considered to achieve an integral view of their social impact.

- Equity, regarding the magnitude of the projects' impact, as well as the social conditions of the benefited individuals, must also be considered. Admittedly, the main difficulty for characterizing the "best public project portfolio" is finding a mechanism to appropriately define, evaluate, and compare social returns. Regardless of the varying definitions of the concept of social return, we can assert the tautological value of the following proposition. Proposition 1: Given two social projects, A and

B, with similar costs and budgets, A should be preferred to B if A has a better social return. Ignoring, for a moment, the difficulties for defining the social return of a project portfolio, given two portfolios, C and D, with equivalent budgets, C should be preferred to D if and only if C has a better social return.

Thus, the problem of searching for the best projectportfolio can be reduced to finding a method for assessing social-project returns, or at least a comparative way to analyze alternative portfolio proposals. The most commonly used method to examine the efficiency impacts of public policies is "cost-benefit" analysis (e.g. Boardman, 1996). Under this approach, the assumed consequences of a project are "translated" into equivalent monetary units where positive consequences are considered "benefits" and negative consequences are considered "loses" or "costs". The temporal distribution of costs and benefits, modeled as net-cash-flows and adjusted by applying a "social discount rate", allows computing the net present-value of individual projects. A positive net present-value indicates that a project should be approved whenever enough resources are available (Fernandez et al., 2009a). Therefore, the net present-value of a particular project can be used to estimate its social return. As a consequence, the social impact of a project portfolio can be computed as the sum of the netpresent-value of all the projects in the portfolio. The best portfolio can then be found by maximizing the aggregated social return (portfolio net-present-social benefit) using 0-1 mathematical programming (e.g. Davis and Mc Keoun, 1986). This cost-benefit approach is inadequate for managing the complex multidimensionality of the combined outcome of many projects, especially when it is necessary to assess intangibles that have no well-defined market values. In extreme cases, this approach favors unacceptable practices (either socially or morally) such as pricing irreversible ecological damages, or even human life. Aside from ethical concerns, setting a price to intangibles for which a market value is highly controversial can hardly be considered a good practice. For a detailed analysis on this issue, the reader is referred to the works by French (1993), Dorfman (1996), and Bouyssou et al. (2000). Despite this drawback, cost-benefit analysis is the preferred method for evaluating social projects (Abdullah and Chandra, 1999). Besides, not using this approach for modeling the multi-attribute impacts of projects leave us with no other method for solving portfolio problems with single objective 0-1 programming. A contending approach to cost-benefit is multi-criteria analysis. This approach encompasses a variety of techniques for exploring the preferences of the Decision Makers (DM), as well as models for analyzing the complexity inherent to real decisions (Fernandez et al., 2009a). Some of the most broadly known multi-criteria approaches are MAUT (cf. Keeney and Raiffa, 1976), AHP (cf. Saaty, 2000, 2005), and outranking methods (Roy, 1990;Figureueira et al., 2005; Brans and Mareschal, 2005). Multi-criteria analysis represents a good alternative to overcome the limitations of costbenefit analysis as it can handle intangibles, ambiguous preferences, and veto conditions. Different multi-criteria methods have been proposed for addressing project evaluation and portfolio selection (e.g. Santhanam and Kyparisis, 1995 ; Badri et al., 2001 ; Fandel and Gal, 2001 ; Lee and Kim, 2001 ; Gabriel et al., 2006; Duarte and Reis, 2006; Bertolini et al., 2006; Mavrotas et al., 2006; Sugrue et al., 2006; Liesio et al., 2007 ; Mavrotas et al., 2008; Fernandez et al., 2009a,b). The advantages of these methods are well documented in the research literature and the reader is referred to Kaplan y Ranjithan (2007) and to Liesio et al. (2007) for an in-depth study on the topic. Multi-criteria analysis offers techniques for selecting the best project or a small set of equivalent "best" projects (this is known as the P_α problem, according to the known classification by Roy (1996)), classifying projects into several predefined categories (e.g. "good", "bad", "acceptable"), known as the P_β problem, and ranking projects according to the preferences or priorities given by the decision maker (the P_γ problem). Given a set of ranked projects, funding resources may be allocated following the priorities implicit in the ranking until no resources are left (e.g. Martino, 1995). This is a simple but rigid process that has been questioned by several authors (e.g. Gabriel et al., 2006, Fernandez et al., 2009 a,b). According to our perspective, the decision on which projects should receive financing must be made based on the best portfolio, rather than on the best individual projects. Therefore, it is insufficient to compare projects to one another. Instead, it is essential to compare portfolios. Selecting a portfolio based on individual projects' ranking guarantees that the set of the best projects will be supported. However, this set of projects does not necessarily equals the best portfolio. In fact, these two sets might be disjoint. Under this scenario, it is reasonable to reject a relatively good (in terms of its social impact) but expensive project if it requires disproportionate funding (Fernandez et al. 2009 a,b). Therefore, obtaining the best portfolio is, we argue, equivalent to solving the $P\alpha$ problem defined over the set of all feasible portfolios. Mavrotas et al. (2008) argue that, when the portfolio is optimized, good projects can be outranked by combinations of low-cost projects with negligible impact. However, this is not a real shortcoming whenever the following conditions are satisfied.

- Each project is individually acceptable
- The decision maker can define his/her preferences over the set of feasible portfolios (by using some quality measure, or even by intuition)
- The decision maker prefers the portfolio composed of more projects with lower costs. In order to solve the selection problem over the set of feasible portfolios, the following issues should be addressed.

- The nature of the decision maker should be defined. It must be clear that this entity can address social interest problems in a legit way. In addition, the following questions should be answered. Is the decision-maker a single person? Or is it a collective with homogeneous preferences such that these can be captured by a decision model? Or is it, instead, a heterogeneous group with conflicting preferences? How is social interest reflected on the decision model?

- A computable model of the DM's preferences on the social impacts of portfolios is required. • Portfolio selection is an optimization problem with exponential complexity. The set of possible portfolios is the power set of the projects applying for funding. The cardinality of the set of portfolios is 2^N, where N is the number of projects. The complexity of this problem increases significantly if we consider that each project can be assigned a support level. That is, projects can be partially supported. Under these conditions, the optimization problem is not only about identifying which projects constitute the best portfolio but also about defining the level of support for each of these projects.

- If effects of synergetic projects or temporal dependencies between them are considered, the complexity of the resulting optimization model increases significantly.

The first issue is related to the concepts of social preferences, collective decision, democracy, and equity. The second issue, on the other hand, constitutes mathematical decision analysis' main area of influence. These capabilities for building preference models that incorporate different criteria and perspectives is what makes these techniques useful (albeit with some limitations) for constructing multidimensional models of conflicting preferences.

The DM's preferences on portfolios (or their social impacts) can be modeled from different perspectives, using different methods, and to achieve different goals. Selecting one of these options depends on who the DM is (e.g., a single person or a heterogeneous group), as well as on how much effort this DM is willing to invest in searching for the solution to the problem. Therefore, the information about the impact and quality of the projects that constitute a portfolio can be obtained from the DM using one of several available alternatives. This requires us to consider different modeling strategies and, in consequence, different approaches for finding the solution to this problem. We should note that the DM's preferences can be modelled using different and varying perspectives; ranging from the normative approach that requires consistency, rationality, and cardinal information, to a totally relaxed approach requiring only ordinal information. The chosen model will depend on the amount of time and effort the decision maker is willing to invest during the modelling process, and on the available information on the preferences. Here, we are interested in constructing a functional-normative model of the DM's preferences on the set of portfolios. Evolutionary algorithms are powerful tools for handling the complexity of the problem (third and fourth issues listed above). Compared with conventional mathematical programming, evolutionary algorithms are less sensitive to the shape of the feasible region, the number of decision variables, and the mathematical properties of the objective function (e.g., continuity, convexity, differentiability, and local extremes). Besides, all these issues are not easily addressed using mathematical programming techniques (Coello, 1999). While evolutionary algorithms are not more time-efficient than mathematical programming, they are often more effective, generally achieving satisfactory solutions to problems that cannot be addressed by conventional methods (Coello et al., 2002). Evolutionary algorithms provide the necessary instruments for handling both the mathematical complexity of the model and the exponential complexity of the problem. In addition, mathematical decision analysis methods are the main tools for modelling the DM's preferences on projects and portfolios, as well as for constructing the optimization model that will be used to find the best portfolio. The rest of this chapter is organized as follows. An overview of the functional-normative approach to decision making, as well as its use as support for solving selection, ranking and evaluation problems is considered in Section 2. In Section 3, we study the public portfolio selection problem where a project's impact is characterized by a project evaluation, and the DM uses a normative approach to find the optimal portfolio (i.e., the case where maximal preferential information is provided). In the same section we also describe an evolutionary algorithm for solving the optimization problem. An illustrative example is provided in Section 4. Finally, some conclusions are presented in Section 5.

An outline of the functional approach for constructing a global preference model

Mathematical decision analysis provides two main approaches for constructing a global preference model using the information provided by an actor involved in a decision-making process. The first of these approaches is a functional model based on the normative axiom of perfect and transitive comparability. The second approach is a relational model better known for its representation of preferences as a fuzzy outranking relation. In this work, however, we will focus on the functional approach only. When using the functional model, also known as the functional-normative approach (e.g. French, 1993), the Decision Maker must establish a weak preference relation, known as the at least as good as relation and represented by the symbol $\succsim$. This relation is a weak order (a complete and transitive relation) on the decision set A. The statement "a is at least as good as b" (a $\succsim$ b) is considered a logical predicate with truth values in the set {False, True}. If a $\succsim$ b is false then b $\succsim$ a must be true, implying a strict preference in favor of b over a. Given the transitivity of this relation, if the DM simultaneously considers that predicates a $\succsim$ b and b $\succsim$ c are true, then, the predicate a $\succsim$ c is also set to true. This approach does not consider the situation where both predicates, a $\succsim$ b and b $\succsim$ a, are false, a condition known as incomparability. Because of this, the functional model requires the DM to have an unlimited power of discrimination. The relation $\succsim$ can be defined over any set whose elements may be compared to each other and, as a result of such comparison, be subject to preferences. Of particular interest is the situation where the decision maker considers risky events and where the consequences of the actions are not deterministic but rather probabilistic. To formally describe this situation, let us introduce the concept of lottery at this point. Definition 1. A lottery is a 2N-tuple of the form $(p_1, x_1; p_2, x_2;... p_N, x_N)$, where $x_i \in R$ represents the consequence of a decision, pi is the probability of such consequence , and the sum of all probabilities equals 1. Given that the relation $\succsim$ is complete and transitive, it can be proven that a real-valued function V can be defined over the decision set A (V: A $\rightarrow$ R), such that for all a, b $\in$ A, V(a) $\geq$ V(b) $\Leftrightarrow$ a $\succsim$ b. This function is known as a value or utility function in risky cases (French, 1993). If the decision is being made over a set of lotteries, the existence of a utility function U can be proven such that $\bar{U}(L_1) \geq \bar{U}(L_2) \Leftrightarrow L_1 \succsim L_2$, where L1 and L2 are two lotteries from the decision set and $\bar{U}$ is the expected value of the utility function (French, 1993). The value, or utility, function represents a well formed aggregation model of preferences. This model is constructed around the set of axioms that define the rational behavior of the decision maker. In consequence, it constitutes a formal construct of an ideal behavior. The task of the analyst is to conciliate the real versus the ideal behavior of the decision maker when constructing this model. Once the model has been created, we have a formal problem definition. This is a selection problem that is solved by maximizing either V or $\bar{U}$ over the set of feasible alternatives. From this, a ranking can be obtained by simply sorting the values of these functions. By dividing the range of these values into M contiguous intervals, discrete ordered categories can be defined for labeling the objects in the decision set A (for instance, Excellent, Very Good, Good, Fair, and Poor). These categories are considered as equivalence classes to which the objects are assigned to. When building a functional model, compatibility with the DM's preferences must be guaranteed. The usual approach is to start with a mathematical formulation that captures the essential characteristics of the problem. Parameters are later added to the model in a way that they reflect the known preferences of the decision maker. Hence, every time the DM indicates a preference for object a over object b, the model (i.e., the value function V)must satisfy condition $V(a) > V(b)$. Otherwise, the model should satisfy condition $V(a) = V(b)$, indicating that the DM has no preference of a over b, nor has the DM a preference of b over a. This situation is known as indifference on the pair (a, b). If V is an elemental function, these preference/ indifference statements on the objects become mathematical expressions that yield the values of V's parameters. To achieve this, usually the DM provides the truth values of several statements between pairs of decision alternatives (a_i, b_i). Then, the model's parameter values are obtained from the set of conditions $V(a_i) = V(b_i)$. Finally, the value and utility functions are generally expressed in either additive or product forms, and, in the most simple cases, as weighted-sum functions. The expected gain in a lottery is the average of the observed gains in the lottery's history. If the DM plays this lottery a sufficiently large number of times, the resulting gain should be close the lottery's expected gain. However, it is not realistic to assume that a DM will face (play) the same decision problem several times as decision problems are, most of the times, unique and unrepeatable. Therefore it is essential to model the DM's behavior towards risk. Persons react differently when facing risky situations. In real life, a DM could be risk prone, risk averse, or even risk neutral. Personal behavior for confronting risk is obviously a subjective characteristic depending on all of the following.

- The DM's personality
- The specific situation of the DM as this determines the impact of failing or succeeding.

    The amount of the gain or loses that will result from making a decision.

- The relationship of the DM with these gains and loses.

    All these aspects are closely related. While the first of them is completely subjective, the remaining three have evident objective features. The ability for modeling the decision maker's behavior when facing risk is one of the most interesting properties of the functional approach. At this point, it is necessary to introduce the concept of certainty equivalence in a lottery. Definition 2. Certainty equivalence is the "prize" that makes an individual indifferent between choosing to participate in a lottery or to receive the prize with certainty. A risk averse DM will assign a lottery a certainty equivalence value lower than the expected value of the lottery. A risk prone DM, on the other hand, will assign the lottery a certainty equivalence value larger than the lottery's expected gain. We say a DM is risk neutral when the certainty equivalence value assigned to a lottery matches the lottery's expected gain. This behavior of the DM yields quite interesting properties on the utility function. For instance, it can be proven that a risk averse utility function is concave, a risk prone utility function is convex, and a risk neutral function is linear. Let us conclude this section by summarizing both the advantages and disadvantages of the functional approach. We start by listing the main advantages of the functionalapproach.

- It is a formal and elegant model of rational decision making.
- Once the model exists, obtaining its prescription is a straight forward process.
- It can model the DM's behavior towards risk.

Now, we provide a list of drawbacks we have identified on the functional approach.

- It cannot incorporate ordinal or qualitative information.
- In real life, DM's do not exactly follow a rational behavior.
- When decisions are made by a collective, the transitivity of the preference relation cannot be guaranteed.
- It cannot precisely model threshold effects, nor can it use imprecise information.
- In most cases, the DM does not have the time to refine the model until a precise utilityfunction is obtained.

A FUNCTIONAL MODEL FOR PUBLIC PORTFOLIO OPTIMIZATION USING GENETIC

algorithms Let us consider a set P_r of public projects whose consequences can be estimated by the DM. These projects have been considered acceptable after some prior evaluation. That is, the DM would support all of them, given that enough funds are available and that no mutually exclusive projects are members of the set. However, projects are not, in general, mutually independent. In fact, they can be redundant or synergetic. Furthermore, they may establish conflicting priorities, or compete for material or human resources, which are indivisible, unique, or scarce. For the sake of generality, let us consider a planning horizon partitioned in T adjacent time intervals. When T=1, this problem is known as the stationary budgeting problem (one budgeting cycle) (Chan et al., 2005). In non-stationary cases, there could be different levels of available funds for each period. In its more general form, a portfolio is a finite set of pairs of projects and periods $\{(p_i, t(pi))\}$, where $p_i \in P_r$ and $t(p_i) \in T$ denotes the period when pi starts. A portfolio is feasible whenever it satisfies financial and scheduling restrictions, including precedence, and it does not contain redundant or mutually exclusive projects. These restrictions may also be influenced by equity, efficiency, geographical distribution, and the priorities imposed by the DM. In particular, if only one budgeting cycle is considered, the portfolios are subsets of Pr. The set of projects is partitioned in different areas, according to their knowledge domain, their social role, or their geographic zone of action. One project can only be assigned to one area. Such partition is usually due to the DM's interest for obtaining a balanced portfolio.Given a set of areas $A = \{A_1, A_2, ..., A_n\}$, the DM can set the minimum and maximum amounts of funding that will be assigned to projects belonging to area Ai $\in$ A.The general problem is to determine which projects should be supported, in what period should the support start, and the amount of funds that each project should receive, provided that the overall social benefit from the portfolio is maximised. In order to have a formal problem statement, we should answer the following questions.

- How can the return of a public project-portfolio be formally defined?
- How can objective and subjective criteria be incorporated for optimizing project- portfolio returns?
- Under what conditions can the return of a portfolio be effectively maximized?
- What methods can be used to select the best portfolio?

To achieve the goal of maximizing social return we need to formally define a real-valuedfunction, Vsocial, that does not contravene the relation $\succsim$social. The construction of such function is, however, problematical due to the following reasons. i. A set of well defined social preferences must exist. ii. This set of preferences must be revealed. The preference-indifference social relation is required to be transitive and complete over social states (premise i). However, due to the known limitations for constructing collective rational-preferences (e.g., Condorcet's Paradox, Arrow's Impossibility Theorem, and context-dependent preferences), (Bouyssou et. al., 2000; Tversky and Simonson, 1993; French, 1993), and to the difficulty in obtaining valid information about social preferences from the decision maker, premises i and ii are rarely fulfilled in real-world cases (Sen, 2000, 2008). The success of public policies is measured in terms of their contribution to social equity and social "efficiency". A project's social impact should be an integrated assessment of such criteria. In the research literature, it is possible to find several methods that have been proposed for estimating a project contribution to social well-fare. Unfortunately, they all show serious limitations for handling intangible attributes. Furthermore, these methods' objectivity for measuring the contribution of each project or public policy is questionable. In any society, a wide variety of interests and ideologies can coexist. This human condition makes it complicated to reach a consensus on what an effective measure of social benefit should be. In turn, the absence of consensus leads to a lack of objectivity on any defined measure. This lack of objectivity is closely related to a nonexistent function of social preference and to the ambiguity of collective preferences as reported by Condorcet, Arrow, and Sen (Bouyssou et al., 2000; Sen, 2000, 2008). While the social impact is objective, its assessment is highly subjective as it depends on the ideology, preferences and values of the person measuring the impact. This subjectivity, however, does not necessarily constitute a drawback as it is not arbitrary. In the end, decision making does not lack of subjective elements. The set of criteria upon which the decision making is based should strive to be objective. However, the assessment of the combined effect of such criteria, some of them in conflict with each other, is subjective in nature as it depends on the perception of the decision maker. The objectivity of decision making theory is not based on eliminating all subjective elements. Instead, it is based on creating a model that reflects the system of values of the decision maker. In every decision problem it is necessary to identify the main actor whose values, priorities, and preferences, are to be satisfied. In this context (the problem of efficiently and effectively allocating public resources), we will call "supra-decision-maker" (SDM) to this single or collective actor. For the rest of the discussion, we drop the idea of modeling public returns from a social perspective in favor of modeling the SDM's preferences. Focusing exclusively on the SDM's preferences is a pragmatic representation of the problem that raises ethical concerns. This is particularly true when the SDM is elected democratically and, as such, his/her decisions formally represent the preferences of the society. In real life, an SDM may possibly have a very personal interpretation of social welfare and subjective parameters to evaluate project returns that do not necessarily represent the generalized social values but rather the ambition of a certain group. Thus, even under the premise ofethical behavior, the SDM –who is supposed to distribute resources according to social preferences– can only act in response to his/her own preferences. The reasons for this are that either the SDM hardly knows the actual social preferences, or he/she pursues his/her own satisfaction –according to his/her preferences– in an honest attempt to achieve what he/she thinks is socially better. Unethical behavior or lack of information can cause the SDM's preferences to significantly deviate from the predominant social interests. In turn, this situation might trigger events such as social protests claiming to reduce the distance between the SDM's preferences and social interests. Therefore, solving a public project- portfolio selection problem is about finding the best solution from the SDM's perspective. This solution (under the premise of ethical behavior) should be close to the portfolio with the highest social return.

A Functional model of the subjective return

In order to maximize the portfolio's subjective return (that is, the return from the SDM perspective), we must build a value function that satisfies relation $\gtrsim_{\text{portfolios}}$. For a starting analogy, let us accept that each project's return can be expressed by a monetary value, in a similar way as cost-benefit analysis. If no synergy and no redundancy exist (or they can be neglected) among the projects, the overall portfolio's return can be calculated as follows.

$$R_t = x_1\, c_1 + x_2\, c_2 + \ldots + x_N\, c_N$$

(1)

In Equation 1, N is the cardinality of P_r. The value of x_i is set to 1 whenever the i-th project is supported, otherwise $x_i = 0$. Finally, c_i is the return value of the i-th project. Let M_i denote the funding requirements for the i-th project. Let d be an N-dimensional vector of real values. Each value, d_i, of vector d is associated to the funding given to the i-th project. If a project is not supported, then the corresponding value in d associated to such project will be set to zero. With this, we can now formally define the problem of portfolio selection.

Problem definition 1. Portfolio selection optimization can be obtained after maximizing R_t, subject to $d \in R_F$, where R_F is a feasible region determined by the available budget, constraints for the kind of projects allowed in the portfolio, social roles, and geographic zones. Problem 1 is a variant of the knapsack problem, which can be efficiently solved using 0-1 programming. Unfortunately, this definition is an unrealistic model for most social portfolio selection problems due to the following issues. 1. For Equation

1. to be valid, the monetary value associated to each project's social impact must be known. Monetary values can be added to produce a meaningfulFigureure. However, due to the existence of indirect as well as intangible effects on such projects, it is unrealistic to assume that such monetary equivalence can be defined for all projects. If we cannot guarantee that every c_i in Equation 1 is a monetary value, then the expression becomes meaningless.

2. Most of the times, the decision is not about accepting or rejecting a project but rather about the feasibility of assigning sufficient funds to it.

3. The effects of synergy between projects can be significant on the portfolio social return. Therefore, they must be modeled. For instance consider the following two projects, one for building a hospital and the other for building a road that will enhance access to such hospital. Both of such projects have, individually, an undeniable positive impact. However their combined social impact is superior.

4.   Time dependences between projects are not considered by Problem definition 1. 5. It is possible that for a pair of projects (i and j) $c_i \gg c_j$ and $M_i \gg M_j$, the solution to this problem indicates that project i should not be supported ($xi = 0$) whereas project j is supported ($x_j = 1$). The SDM might not agree to this solution, as it fails to support a high-impact project while it provides funds to a much less important project. Furthermore, such situation will be difficult to explain to the public opinion.

The functional normative approach presented in Section 2 is used to address the first issue on this list. Here, we present a new approach based on the work of Fernandez and Navarro (2002), Navarro (2005), Fernandez and Navarro (2005), and Fernandez et al. (2009). Addressing issues 2 to 5 on the list above requires using a heuristic search and optimization methods.

This new approach is constructed upon the following assumptions. Assumption 1: Every project has an associated value subjectively assigned by the SDM. This value increases along with the project's impact. Assumption 2: This subjective value reflects the priority that the SDM assigns to the project. Each project is assigned to a category from a set of classes sorted in increasing order of preference. These categories can be expressed qualitatively (e.g., {poor, fair, good, very good, excellent}) or numerically in a monotonically increasing scale of preferences. Assumption 3: Projects assigned to the same category have about the same subjective value to the SDM. Therefore, the granularity of the discrete scale must be sufficiently fine so that no two projects are assigned to the same class if the SMD can establish a strict preference between them. Assumption 4 (Additivity): The sum of the subjective values of the projects belonging to a portfolio is an ordinal-valued function that satisfies relation $\gtrsim_{portfolios}$. Fernandez et al. (2009) rationalize this last assumption by considering that each project is a lottery. A portfolio is, in consequence, a "giant" lottery being played by a risk-neutral SDM. Under this scenario, the subjective value of projects and portfolios corresponds to their certainty equivalent value. Under Assumption 4, the interaction between projects cannot be modeled. Synergy and redundancy in the set of projects are characteristics that require special consideration that will be introduced later. Under Assumptions 1 and 4, the SDM assess a subjective value to portfolio given by the following equation.

$$V = x_1 c_1 + x_2 c_2 + \ldots + x_N c_N$$

(2)

In Equation 2, c_i represents the subjective value of the i-th project. Equations 1 and 2 are formally equivalent. However, the resulting value of V only makes sense if there is a process to assign meaningful values to ci. Before we proceed to the description of the rest of the assumptions, we need to introduce the concept of elementary portfolio. Definition 3: An elementary portfolio is a portfolio that contains only projects of the same category. It will be expressed in the form of a C-dimensional vector, where C is the number of discrete categories. Each dimension is associated to one particular category. The value in each dimension corresponds to the number of projects in the associated category. Consequently, the C-dimensional vector of an elementary portfolio with n projects will have the form $(0, 0, ..., n, 0, ..., 0)$. Assumption 5: The SDM can define a complete relation $\succsim$ on the set of elementary portfolios. That is, for any pair of elementary portfolios, P and Q, one and only one of the following propositions is true.

- Portfolio P is preferred to portfolio Q
- Portfolio Q is preferred to portfolio P
- Portfolios P and Q are indifferent.

Assumption 6 (Essentiality): Given two elementary portfolios, P and Q, defined over the same category. Let $P = (0, 0, ..., n, 0, ..., 0)$ and $Q = (0, 0, ..., m, 0, ..., 0)$. P is preferred to Q if an only if $n > m$. From the set of discrete categories, let C_1 be the lowest category, CL be the highest, and C_j a category preferred to C_1. Assumption 7 (Archimedean): For any category C_j, there is always an integer value n such that the SDM would prefer a portfolio composed of n projects in the C_1 category to any portfolio composed of a single project in the C_j category.

Assumption 8 (Continuity): If an elementary portfolio $P = (x, 0, ..., 0, ..., 0)$ is preferred to an elementary portfolio $Q = (0, ..., 1, 0, ..., 0)$, defined over category j for $1 < j \le L$, there is always a pair of integers values n and m ($n > m$) such that an elementary portfolio with n projects of the lowest category is indifferent to another elementary portfolio with m projects of the j-th category. Assumption 5 characterizes the normative claim of the functional approach for decision making. Assumption 6 is a consequence of Assumption 4 (additivity) combined with the premise that all projects satisfy minimal acceptability requirements. Assumption 7 is a consequence of both essentiality and the non-bounded character of the set of natural numbers. Assumption 8 simulates the way in which a person balances a scale using a set of two types of weights whose values are relative primes. Let us say that c_1 is a number representing the subjective value of the projects belonging to the lower category C_1. Similarly, let us use c_j to represent the value of projects in category C_j. Now, suppose that the elementary portfolios P (containing n projects in C_1) and Q (integrated by m projects in C_j) are indifferent. That is, P and Q have the same V value. If we combine Assumption 8 with Equation 2, we obtain the following expression.

$$n\, c_1 = m\, c_j \Leftrightarrow c_j = (n/m)c_1$$

If V is a value function, then every proportional function is also a value function satisfying the same preferences. Therefore, we can arbitrarily set $c_1=1$ to obtain Equation 3 below.

$$c_j = n/m \qquad (3)$$

In consequence, Equation 2 can now be re-stated as follows.

$$U = \Sigma_{i,k}\, w_{ik} x_{ik} \qquad (4)$$

In Equation 4, the variable j is used to index categories, whereas variable k indexes projects. The value of w_{1k} is set to 1, and $w_{jk} = n/m_j$, where m_j denotes the cardinality of an elementary portfolio defined over category C_j. Additionally, factors w_{ik} might be interpreted as importance factors. These weights express the importance given by the SDM to projects within certain category. Therefore, they should be calculated from the SDM's preferences, expressed while solving the indifference equations between elementary portfolios, as stated by Assumption 8 and according to Equation 3. A weight must be calculated for every category. If the cardinality of the set of categories is too large, the resolution of such categories can be reduced to simplify the model. A temporary set of weights is obtained using these coarse categories. By interpolation on such set, the values of the original (finer resolution) set can be obtained.

Fuzziness of requirements

Another important issue is the imprecise estimation of the monetary resources required by each project. If d_k are the funds assigned to the k-th project, then there is an interval $[m_k, M_k]$ for d_k where the SDM is uncertain about whether or not the project is being adequately supported. Therefore, the proposition "the k-th project is adequately supported" may be seen as a fuzzy statement. If we consider that the set of projects with adequate funds is fuzzy, then the SDM can define a membership function $\mu_k(d_k)$ representing the degree of truth of the previous proposition. This is a monotonically increasing function on the interval

$[m_k, M_k]$, such that $\mu_k(M_k) = 1$, $\mu_k(m_k) > 0$, and $\mu_k(d_k) = 0$ when $d_k < m_k$.

The subjective value assigned by the SDM to the k-th project is based on the belief that the project receives the necessary funding for its operation. When $d_k < M_k$, the SDM hesitates about the truth of that statement. This uncertainty affects the subjective value of the project, because it reduces the feasible impact of the project, which had been subjectively estimated under the premise that funding was sufficient. The reduction of the project's subjective value can be modeled by the product of the original value and a feasibility factor f. This factor is a monotonically increasing function with μ_k as an argument such that $f(0) = 0$ and $f(1) = 1$. Equation 5 below, is generated by introducing this factor into Equation 4, and assuming that

$$f(\mu_{ik}) > 0 \Leftrightarrow x_{ik} = 1.$$

$$U = \Sigma_{ik} \, f(\mu_{ik}) \, w_{ik}$$

(5)

The simplest definition of the feasibility factor is to make $f(\mu_{ik}) = \mu_{ik}$. This is equivalent to a fuzzy generalization of Equation 4. In such case, x_{ik} can be considered as the indicator function of the set of supported projects. When a non-fuzzy model includes the binary indicator function of a crisp set, the fuzzy generalization provided by classical "fuzzy technology" is made substituting this function with a membership function expressing "the degree of membership" to the more general fuzzy set. In this way, Equation 5 becomes Equation 6 shown below.

$$U = \Sigma_{ik} \, w_{ik} \, \mu_{ik}$$

(6)

Equation 6 was proposed by Fernandez and Navarro (2002) as a measure of a portfolio's subjective value.

Synergy and redundancy Redundancy between projects can be addressed using constraints. For every pair of redundant projects, (p_i, p_j), $i < j$, condition $\mu_i(d_i) \times \mu_j(d_j) = 0$ should be enforced. Let $S = \{S_1, S_2, ..., S_k\}$ be the set of coalitions of synergetic projects. In a model like the one represented by Equation 5, each of these coalitions should be treated as an (additional) individual project. As a result, each coalition has an associated cost (i.e., the sum of the costs of the individual projects in the coalition), and an evaluation. This evaluation should be better than the evaluation of any of the projects in the coalition. Let us assume that coalitions S_i and S_j become projects P_{N+i} and P_{N+j}, respectively. If S_i is a subset of S_j then it does not make sense to include them both in a portfolio. Therefore, P_{N+i} and P_{N+j} must be considered redundant projects. Furthermore, if project p_n is a member of S_j, then the pair $(p_n, p_{N+}i)$ is also redundant (since the value of p_n is included in the value of p_{N+i}).

Genetic algorithm for optimizing public portfolio subjective value

Suppose that a feasible region of portfolios, RF, is defined by constraints on the total budget and on the distribution of projects by area. In addition, the SDM could include further constraints on the portfolios due to following reasons.

- The particular budget distribution of the portfolio could be very difficult to justify. Let us suppose that the SDM asserts that "project p_j is much better than project p_i". In consequence, any portfolio in which μi is greater than μj could be unacceptable. This implies the existence of some veto situations that can be modeled with the following constraint. For every project pi and p_j, being s_i, and s_j their corresponding evaluations, if $(s_i - s_j) \geq v_s$, then $(\mu_i(d_i) - \mu j(d_j))$ must be greater than (or equal to) 0, where vs is a veto threshold. In the following they will be called veto constraints.

- A possible redundancy exists between projects. Let us use R'_F, $R'_F \subset R_F$, to denote the set of values for the decision variables that make every portfolio acceptable. All the veto constraints are satisfied in R'_F and there are no redundant projects in the portfolios belonging to this region. The optimization problem can now be defined as follows.

Problem definition 2. An optimal portfolio can be selected by maximizing $U = \Sigma_{ik} f(\mu_{ik}(d_{ik}))\ w_{ik}$, subject to $d \in R'F$, where d_{ik} indicates the financial support assigned to the k-th project belonging to the i-th category

Solving this problem requires a complex non linear programming algorithm. The number of decision variables involved can be in the order of thousands. Due to the discontinuity of μ_i the objective function is discontinuous on the hyper planes defined by $d_{ik} = m_{ik}$. Therefore, its continuity domain is not connected. The shape of the feasible region $R'F$ is too convoluted, even more if synergy and redundancy need to be addressed. R'_F hardly has the mathematical properties generally required by non linear programming methods. Note that veto constraints on the pairs of projects (p_i, p_k) and (p_j, p_k') are discontinuous on the hyper planes defined by $d_{ik} = m_{ik}$ and $d_{jk}' = m_{jk}'$. In a real world scenario, where hundreds or even thousands of projects are considered, non-linear programming solutions cannot handle these situations. Using Equation 6, a simplified form of Problem definition 2, was efficiently solved by Fernandez et al. (2009) and later by Litvinchev et al. (2010) using an integer-mixed programming model. Unfortunately, this approach cannot handle synergy, redundancy, veto constraints, nor can it handle the non-linear forms of function f in Problem definition 2. Evolutionary algorithms are less sensitive to the shape of the feasible region, the number of decision variables, and the mathematical properties of the objective function (e.g., continuity, convexity, differentiability, and local extremes). In contrast, all of these issues are a real concern for mathematical non linear programming techniques (Coello, 1999). While evolutionary algorithms are not time-efficient, they often find solutions that closely approximate the optimal. Problem definition 2 represents a relatively rough model. However, the main interest is not on fine tuning the optimization process but rather on the generality of the model and on the ability to reach the optimal solution or a close approximation. InFigureure 1, we illustrate the genetic algorithm used for solving the optimization problem stated in Problem definition 2. This algorithm is based on the work of Fernandez and Navarro (2005). As in any genetic algorithm, a fundamental issue is defining a codification for the set of feasible solutions to the optimization problem. In this case, each individual represents a portfolio and each chromosome contains N genes, where N is the number of projects. For the chromosome, we use a floating point encoding representing the distribution of funding among the set of projects in the portfolio. The financial support for each project is represented by its membership function, $\mu_j(d_j)$, which is realvalued with range in [0, 1]. That is, a floating point number represents each project's membership value. This membership value is a gene in our definition of chromosomes. As discussed earlier, the number of genes can be increased in order to address the effects of synergetic projects. The fitness value of each individual is calculated based on function U given by Equation 5. Remember that this is a subjective value that captures the SDM's certainty that the project receives the necessary funding for its operation. The SDM's idea that a project has been assigned sufficient funds is modeled using two parameters, α and β. The domain for both parameters is the continuous interval [0, 1].The first parameter, α, can be interpreted as the degree of truth of the assertion "the project has sufficient financial support if it receives m monetary units of funding". When this financial support reaches the value βM, the predicate "the project has sufficient funding" is considered true. The value of these two parameters is needed to establish models for function μj in order to calculate the value of U. To generate these models, we propose to choose parameters α and β ($0 < \alpha < \beta \leq 1$) for modelling μ as shown inFigureure 2. For the experiments presented here, the values of $\alpha = 0.5$ and $\beta = 1$ have been used. The most promising values for these parameters are reasonably found in the intervals [0.5, 0.7] and [0.9, 1], respectively.

Algorithm 1. A Genetic Algorithm for Project Portfolio Selection.	
Input:	cycles, the number of iterations before the algorithm converges
	generations
	c_r, the Crossover rate
	m_r, the Mutation rate
Output:	best_solution, the best solution found
1	Set best_solution ← any feasible portfolio. // *this is the best so far*
2	Set N ← the number of projects (chromosomes)
3	Set Population ← {best_solution}
4	**for** (i = 1 to cycles) **do**
5	**for** (j = 1 to N - 1) **do**
6	set new_solution ← best_solution
7	randomly select a gene in new_solution and mutate it
8	set Population ← Population ∪ {new_solution}
9	**end**
10	evaluate every individual ∈ Population
11	set best_solution ← the fittest individual ∈ Population
12	**for** (k = 1 to generations) **do**
13	perform crossover on (N × c_r) individuals ∈ Population
14	perform mutation on (N × m_r) individuals ∈ Population
15	set Population ← Population ∪ {best_solution}
16	evaluate every individual ∈ Population
17	set best_solution ← the fittest individual ∈ Population
18	**end**
19	**end**
20	**return** best_solution

Figure. 1: A Genetic Algorithm for Project Portfolio Selection

For the selection stage, the roulette wheel technique was used. That is, the probability that a particular individual is selected for reproduction is proportional to its fitness value. For the experiments, the crossover rate was set to 0.2. Therefore twenty percent of the population is selected for crossover in any given reproductive trial. The crossover operator takes genes from each parent string and combines them to produce the offspring of the next generation. The main reason for doing this is that by creating new strings from fit parent strings, new and promising zones of the search space will be explored. While many crossover techniques have been reported, in this algorithm the classic crossover technique based on a random cut point was used. The number of offspring resulting from this process is one fifth the size of the population. The replacing process dictates how to update the current population with the individuals obtained by crossover.

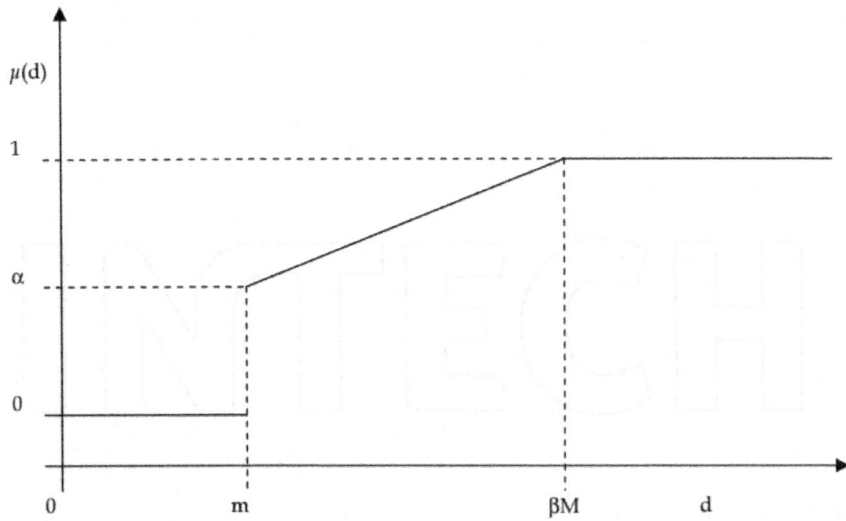

Figure. 2: The Membership Function

A random replacement approach (every individual has the same probability to be replaced) is used for reducing selective pressure. A similar approach is used for implementing an elitist policy. That is, an individual is randomly chosen from the current population and is replaced by the individual with the highest evaluation. Consequently, the presence of the best individual (best_solution in Algorithm 1) in the updated population is guaranteed. Algorithm 1 uses a constant mutation rate that is set a priori. Each individual in the population is considered for mutation, and all the individuals have the same probability of mutating, which is defined by the mutation rate. Once an individual has been selected for mutation, one of its genes is randomly chosen. This gene will change by adding to it a random value in the [-0.2, 0.2] interval, excluding zero. The resulting gene value, however is limited to the [0, 1] interval. Redundancy is addressed in a very simple way. If, as the result of some genetic operator an individual (i.e., a portfolio) containing redundant projects is generated, this individual is immediately "killed". That is, its incorporation to the current population is denied.

An illustrative example

Let us now consider the following example taken from (Fernandez and Navarro, 2005). The goal is to distribute a budget of 50 million dollars among of 400 R&D projects. These projects are distributed in four areas, namely engineering, life sciences, formal sciences, and social sciences. There are 140 projects in the first area (engineering), 80 projects in the second one (life sciences), 100 projects in the third area (formal sciences), and 80 project in the last area (social sciences). No synergetic effects are considered. The classification of the projects, according to their evaluations and areas, is described in Table 1. The projects subjective values corresponding to each category and area are shown in Table 2. These values were obtained taking a social sciences project evaluated as Below Average as baseline ($w = 1$). These values define a ranking on the set of projects that can be used to allocate funds according to the conventional heuristic described in Section 1 (with all its known limitations).

Table 1: Distribution of Projects by Area

	Area 1	Area 2	Area 3	Area 4
Very Good	54	28	13	12
Good	23	9	18	24
Above Average	62	32	36	28
Average	1	9	17	11
Below Average	0	2	16	5
Total	**140**	**80**	**100**	**80**

Table 2: Projects Subjective Values

	Area 1	Area 2	Area 3	Area 4
Very Good	5.838	4.3785	3.892	2.9190
Good	4.540	3.4055	3.027	2.2700
Above Average	3.027	2.2700	2.018	1.5135
Average	2.108	1.5810	1.405	1.0540
Below Average	2.000	1.5000	1.333	1.0000

Four different instances of the problem were generated by assigning random budget ranges to each area. For each project, random values of m_{ik}, and M_{ik} were defined, representing its minimum and maximum funding requirements. The proposed evolutionary algorithm was run 30 times to optimize the expression given by Problem definition 2. For simplicity $f(\mu_{ik})$ was taken to be identical to μ_{ik}

The algorithm was coded using Visual C++. Its execution time was about 25 minutes for one million generations running on a Pentium-4 processor with a, 2.1 GHz clock cycle. This architecture was complemented with 256 MB of physical memory and a 74.5-GB hard disk drive. The experimental results shown in Table 3 indicate a significant improvement in the value of the optimized portfolio with respect to conventional approaches. These results represent an average saving of 6.514 million dollars, equivalent to 13.02% of the total budget. This improvement has a positive impact on the number of supported projects, as Table 4 reveals. The average number of supported projects is 12.5 % higher than when conventional methods were used.

Table 3: Traditional Funding versus our Approach

Instance	Value of the portfolio funding following the ranking given by project evaluations	Value of the optimized portfolio	Improvement
1	1406.80	1533.95	9%
2	1282.36	1496.16	16.67%
3	1279.58	1458.48	14%
4	1393.58	1566.97	12.44%

Table 4: Traditional Funding versus our Approach (portfolio's cardinality)

Instance	Number of supported projects funding following the ranking given by project evaluations	Number of supported projects in the optimized portfolio	Increment
1	237	267	12.76%
2	257	285	10.89%
3	265	299	12.83%
4	246	279	13.41%

Modeling temporal dependencies

The model described in Problem definition 2 can be generalized to incorporate temporal restrictions. Problem definition 3. An optimal portfolio of projects with temporal dependencies can be selected by maximizing $U = \Sigma_{ik} \, f(\mu_{ik}(d_{ik})) \, w_{ik}$, subject to $(\mathbf{d}, \mathbf{t}) \in R''_F$, where vector $\mathbf{t} = (t(p_1), \quad t(p_2), \ldots)$ denotes the decision variables valid during the period of time when each project starts. $R"_F$ contemplates time-precedence restrictions, restrictions on the time projects can start, and the available funds for each time interval.

This problem can be solved using a genetic algorithm similar to the one previously presented. However, a different encoding for individuals must be devised. Our proposal is to encode individuals as a 2N-dimensional vector of the form $(\mu_1, t_1, \mu_2, t_2, ..., \mu_N, t_N)$. As before, genes corresponding to μi have domain defined by the continuous interval $[0, 1]$. Genes corresponding to t_i have a domain defined by the set $\{1, 2, 3, ..., T\}$, where T is the maximum number of time periods. Crossover can only occur between genes of the same kind. However, mutations may occur at any gene. Restrictions such as time precedence and the earliest time a project can start are controlled by constraints as described by Carazo et al. (2010).

CONCLUDING REMARKS

Given a set of premises, it is possible to create a value model for selecting optimal portfolios from an SDM perspective. While this problem is Turing-decidable, finding its exact solution requires exponential time. However, the use of genetic algorithms for solving this problem can closely approximate the optimal portfolio selection. Inspired by a normative approach, the set of premises presented here is based on the following assumptions.

- To the SMD, every project and every portfolio has a subjective value that depends on its social impact. This value exists even if it cannot be initially quantified.

- The SDM either has already defined a consistent system of preferences, or has the aspiration of doing so.

- The SDM is willing to invest a considerable amount of mental effort in order to define this consistent set of preferences and produce the aforementioned value model. As for the algorithmic solution to the portfolio problem, its computational complexity can increase considerably when synergic effects and temporal dependencies are considered. However strategic planning requires a high quality model. The problems defined in this scenario are so important that they justify the use of computational intensive solutions.

ACKNOWLEDGEMENTS

This work was sponsored in part by the Mexican Council for Science and Technology (CONACyT) under grants 57255 and 106098.

REFERENCES

1. Abdullah, A. & Chandra, C.K. (1999). Sustainable Transport: Priorities for Policy Sector Reform, World Bank, Retrieved from

2. Badri, M.A. & Davis, D. (2001). A Comprehensive 0-1 Goal Programming Model for Project Selection. International Journal of Project Management, No. 19, pp. 243-252.

3. Bertolini, M., Braglia, M., & Carmignani, G. (2006). Application of the AHP Methodology in Making a Proposal for a Public Work Contract. International Journal of Project Management No. 24, pp. 422-430.

4. Boardman, A. (1996). Cost-benefit Analysis: Concepts and Practices, Prentice Hall.

5. Boyssou, D., Marchant, Th., Perny, P., Tsoukias, A., & Vincke, Ph. (2000). Evaluations and Decision Models: A Critical Perspective, Kluwer Academic Publishers, Dordrecht. Brans, J.P. & Mareschal, B. (2005).

6. PROMETHEE Methods, In: Multiple Criteria Decision Analysis: State of the Art Surveys, Figureueira, J., Greco, S., & Erghott, M., pp. 163-190, Springer Science + Business Media, New York.

7. Carazo, A.F., Gomez, T., Molina, J., Hernandez-Diaz, A.G., Guerrero, F.M., & Caballero, R. (2010). Solving a Comprehensive Model for Multiobjective Portfolio Selection. Computers & Operations Research No. 37, pp. 630-639.

8. Chan, Y., DiSalvo, J., & Garrambone, M., A. (2005). Goal-seeking Approach to Capital Budgeting. Socio-Economic Planning Sciences, No.39, pp. 165-182.

9. Coello, C. (1999). A Comprehensive Survey of Evolutionary-based Multiobjective Optimization Techniques. Knowledge and Information Systems, No. 1, pp. 269-308

10. Coello, C., Van Veldhuizen, D., & Lamont, G. (2002). Evolutionary Algorithms for Solving Multi-objective Problems, Kluwer Academic Publishers, New York-BostonDordrecht-London-Moscow.

11. Davis, K. & Mc Kewon, P. (1986). Quantitative Models for Management (in Spanish), Grupo Editorial Iberoamérica, Mexico. Dorfman, R. (1996). Why Cost-benefit Analysis is Widely

12. Disregarded and what to do About It?, Interface, Vol. 26, No. 1, pp. 1-6.

13. Duarte, B. & Reis, A. (2006). Developing a Projects Evaluation System Based on Multiple Attribute Value Theory. Computers & Operations Research, No. 33, pp. 1488-1504.

14. Fandel, G. & Gal, T. (2001). Redistribution of Funds for Teaching and Research among Universities: The Case of North Rhine Westphalia. European Journal of Operational Research, No. 130, pp. 111-120.

15. Fernandez, E. & Navarro J. (2002). A Genetic Search for Exploiting a Fuzzy Preference Model of Portfolio Problems with Public Projects, Annals of Operations Research, No. 117, pp. 191-213.

16. Fernandez, E. & Navarro J. (2005). Computer–based Decision Models for R&D Project Selection in Public Organizations. Foundations of Computing and Decision Sciences, Vol. 30, No.2, pp. 103-131.

17. Fernandez, E., Felix, F., & Mazcorro, G. (2009). Multiobjective Optimization of an Outranking Model for Public Resources Allocation on Competing Projects. International Journal of Operational Research, No. 5, pp. 190-210.

18. Fernandez, E., Lopez, F., Navarro, J., Litvinchev, I., & Vega, I. (2009). An Integrated Mathematical-computer Approach for R&D Project Selection in Large Public Organizations. International Journal of Mathematics in Operational Research, No. 1, pp. 372-396.

19. Figureueira, J., Greco, S., Roy, B., & Słowiński, R. (2010). ELECTRE Methods: Main Features and Recent Developments, In: Handbook of Multicriteria Analysis, Applied Optimization, Zopounidis, C., & Pardalos, M.., pp. 51-89,

20. Springer, Heidelberger-DordrechtLondon-New York. French, S. (1993). Decision Theory: An Introduction to the Mathematics of Rationality, Ellis Horwood, London.

21. Gabriel, S., Kumar, S., Ordoñez, J., & Nasserian, A. (2006). A Multiobjective Optimization Model for Project Selection with Probabilistic Consideration. Socio-Economic Planning Sciences, No. 40, pp. 297-313. Kaplan, P. & Ranjithan, S.R., (2007)

22. A new MCDM Approach to Solve Public Sector Planning Problems, Proceedings of the 2007 IEEE Symposium on Computational Intelligence in Multi Criteria Decision Making, pp. 153-159.

23. Gabriel, S., Kumar, S., Ordoñez, J., & Nasserian, A. (2006). A Multiobjective Optimization Model for Project Selection with Probabilistic Consideration. Socio-Economic Planning Sciences, No. 40, pp. 297-313. Kaplan, P. & Ranjithan, S.R., (2007).

24. A new MCDM Approach to Solve Public Sector Planning Problems, Proceedings of the 2007 IEEE Symposium on Computational Intelligence in Multi Criteria Decision Making, pp. 153-159.

25. Liesio, J., Mild, P., & Salo, A. (2007). Preference Programming for Robust Portfolio Modeling and Project Selection. European Journal of Operational Research, No. 181, pp. 1488- 1505.

26. Litvinchev, I., Lopez, F., Alvarez, A., & Fernandez, E. (2010). Large Scale Public R&D Portfolio Selection by Maximizing a Biobjective Impact Measure, IEEE Transactions on Systems, Man and Cybernetics , No. 40, pp. 572-582.

27. Martino, J. (1995). Research and Development Project Selection, Wiley, NY- Chichester-BrisbaneToronto-Singapore.

28. Mavrotas, G., Diakoulaki, D., & Caloghirou, Y. (2006). Project Prioritization under Policy Restrictions. A combination of MCDA with 0-1 Programming. European Journal of Operational Research, No.171, pp. 296-308.

29. Mavrotas, G., Diakoulaki, & D., Koutentsis, A. (2008). Selection among Ranked Projects under Segmentation, Policy and Logical Constraints. European Journal of Operational Research, No. 187, pp. 177-192

30. . Navarro, J. (2005). Intelligent Techniques for R&D Project Selection in Public Organizations (in Spanish), PhD. Dissertation, Autonomous University of Sinaloa, Mexico.

31. Peterson, S. (2005). Interview on Financial Reforms in Developing Countries, Kennedy School Insight, John Kennedy School of Government, Harvard University, Retrieved from: Roy, B. (1990).

32. The Outranking Approach and the Foundations of ELECTRE Methods, In: Reading in Multiple Criteria Decision Aid, Bana and Costa, C.A., pp. 155-183, Springer-Verlag, Berlin,.

33. Roy, B. (1996). Multicriteria Methodology for Decision Aiding, Kluwer. Saaty, T. L. (2000). Fundamentals of the Analytic Hierarchy Process, RWS Publications, Pittsburg.

34. Saaty, T. L. (2005). The Analytic Hierarchy and Analytic Network Processes for the Measurement of Intangible Criteria for Decision-making, In: Multiple Criteria Decision Analysis: State of the Art Surveys,Figureueira, J., Greco, S. and Erghott, M., pp. 345-407, Springer Science + Business Media, New York

35. Santhanam, R. & Kyparisis, J. (1995). A Multiple Criteria Decision Model for Information System Project Selection. Computers and Operations Research, No. 22, pp.807-818.

36. Sen, A. (2000). Development as Freedom, Anchor Books, New York. Sen, A. (2008). On Ethics and Economics (18th Edition), Blackwell Publishing, Malden-OxfordCarlton.

37. Sugrue, P., Mehrotra, A., & Orehovec, P.M. (2006). Financial Aid Management: An Optimization Approach. International Journal of Operational Research, No. 1 pp. 267- 282.

38. Tversky A. & Simonson I. (1993). Context Dependent Preferences. Management Science, No. 39, pp. 1179-1189.

Chapter 4

THE SEARCH FOR PARAMETERS AND SOLUTIONS: APPLYING GENETIC ALGORITHMS ON ASTRONOMY AND ENGINEERING

Annibal Hetem Jr.

Universidade Federal do ABC Brasil

INTRODUCTION

Genetic Algorithms (GAs) can help solving a great variety of complex problems, and the characterization of these problems as possible subject for GA is the first step in applying this technique. After some years, we have used this strong tool to solve problems from astronomy and engineering, and both fields demand complex models and simulations. With the aim of improving previous models and test new ones, we have developed a methodology generate solutions based on GAs. From a first analysis, one must establish the model input and output parameters, and then workout on the inversion of the problem, what we called the inverted model. This concept leads to the final formalism that can be subject to the GA implementation. After a brief presentation of the main concerns and ideas, it will be described some applications and their results and discussions. Some details on implementation are also given together with the particularities of each model/solution. A special section regarding error bars estimates is also provided. The GA method gives a good quality of fit, but the range of input parameters must be chosen with caution, as unrealistic parameters can be derived. GAs can also be used to verify if a given model is better than another for solving a problem. Even considering the limitation of the derived parameters, the automatic fitting process provides an interesting tool for the statistical analysis large samples of data and the models considered.

CHARACTERIZATION OF NP-COMPLETE PROBLEMS

In this section, the NP-Complete problems are presented as the main targets of GAs. Before starting to project a GA, it is of greatest importance to study and characterize the problem to justify the technique to use. The early first notion of NP-completeness was proposed by Stephen Cook (1971), in his famous paper

The complexity of theorem proving procedures. The main ideas presented in this section have their origins in the excelent works of Garey & Johnson (1979) and Papadimitriu (1995).

Deep inside any GA code there is a model of the inverted problem to be solved. This routine works like I don't know what the correct answer is, but I kwon if a candidate to an answer is good or bad. So, the problem to be solved by a GA must have the property that any proposed solution to an instance must be quickly checked for correctness. For one thing, the solution must be concise, with length polynomially bounded by that of the instance. To formalize the notion of quick checking, we will say that there is a polynomial-time algorithm that takes as input instance and the solution and decides whether or not it is a solution. If a problem demands a nondeterministic polynomial time to be solved, it is said a NP-problem, as defined by complexity theory researchers. It means that a solution to any search problem can be found and verified in polynomial time by nondeterministic algorithm.

INVERTING THE PROBLEM

The most remarkable characteristic of a NP-complete problem is the lack known algorithms to find its solution. In a P-Problem, any given candidate to solution can be verified quickly for its accuracy or validity. On the other hand, the time required to solve a NP-problem using any currently known search algorithm increases exponentially with the size of the problem grows. As a consequence, one of the principal unsolved problems in computer science today is determining whether or not it is possible to solve these problems quickly, called the P versus NP problem.

Then, suppose one has a problem M to be solved and asks if a GA based program could solve it. The steps to be followed are:

1. To write down formally the set of parameters to be found, something like $S=\{p_1, p_2, p_3, ..., p_n\}$, where the pi set is a representation of the input parameters. Each pi must be a single number (float or integer), so the S set could be interpreted as a chromosome and each pi as a gene.

2. To express the problem as a function of the set of parameters: $M=f(S)$, with $M=\{q_1, q_2, q_3, ..., q_m\}$, where the qi set is the representation of the output (desired) parameters.

3. Obtain the inverse problem, or the formalities need to compute $S= g(M) =f^{-1}(M)$.

If the g(M) function can be translated to a writable algorithm, and this algorithm is computable in a finite time, then the g(M) is a P-problem. If the f(S) function cannot be translated to a writable algorithm, or this algorithm is computable only with by verifying all possibilities in the S space, then the f(S) is a NP-problem. With both answers: the f(S) function is a NP-problem, and its inverse, g(M) is a P-problem, then the problem can be solved by a GA.

APPLICATIONS ON ASTROPHYSICS

Astrophysics is a field of research very rich in NP-complete problems. Many of actual astrophysicists deal with non-linear systems and unstable conditions. In some cases, the comparative data, or the environment in GA jargon, is an image originated in telescopes or instruments placed in deep space. It is common the need for fit models with multi-spectral

data, like radio, infrared, visible and gamma-rays. All these solution constraints lead to an incredible variety of possibilities for using GA tools. In this section, it will be presented how GAs were used to model protoplanetary discs, an application that involves non-linear radiative-density profile relations. The model combines spectral energy distribution, observed in a wide range of the electromagnetic spectrum, and emissivity behaviour of different dust grain species. Another interesting application is the use of GAs together with and spectral synthesis in the calculation of abundances and metallicities of T Tauri stars. In this problem, the model is outside the GA code, as one of the conditions imposed is to use a standard, well tested, spectral generator. It is presented how to deal with the challenge of changing a ready to use tool into a NP-complete problem and invert it.

Using GA to model protoplanetary discs

This subsection is based on the published work The use of genetic algorithms to model protoplanetary discs (Hetem & Gregorio-Hetem 2007). During its formation process, a young star object (YSO) can be surrounded by gas, dust grains and debris, that shall be gravitationally (and also electrostatically) agglomerate in the future solar system bodies. This material receives the energy brought from the star surface and re-irradiates it in other wavelengths. The contribution of this circumstellar matter to the spectral energy distribution (SED) slope is often used to recognize different categories of young YSOs by following an observational classification based on the near-infrared spectral index (Lada & Wilking 1984; Wilking, Lada & Young 1989; André, Ward-Thompson & Barsony 1993). Actually, this classification suggests a scenario for the evolution of YSOs, from Class 0 to Class III, which is well established for TTs. Here, the adopted model is a flared configuration, according to

Dullemond et al. (2001) modelling of a passively irradiated circumstellar disc with an inner hole. We used this model as the P-problem core of a GA based optimization method to estimate the circumstellar parameters

Presenting the problem

In this subsection we describe the implementation of the GA method for the flared-disc model. The SED for a given set of parameters is evaluated according to Dullemond et al. (2001) model equations. The disc is composed by three components: the inner rim, the shadowed region, and the flared region with two layers: an illuminated hot layer and an inner cold layer. The disc parameters are: radius, RD; mass, MD; inclination, θ; density power law index, p; and inner rim temperature, T_{rim}. The stellar parameters are: distance, d; mass, $M_\star$; luminosity, $L_\star$; and temperature, $T_\star$. The model starts by establishing a vertical boundary irradiated directly by the star, which considers the effect arising from shadowing from the rim, and the variations in scale height as a function of the radius. Figure 1 presents the obtained SED for the star AB Aurigae, as presented in Hetem & Gregorio-Hetem (2007) The model starts by establishing a vertical boundary irradiated directly by the star, which considers the effect arising from shadowing from the rim, and the variations in scale height as a function of the radius. Figure 1 presents the obtained SED for the star AB Aurigae, as presented in Hetem & Gregorio-Hetem (2007).

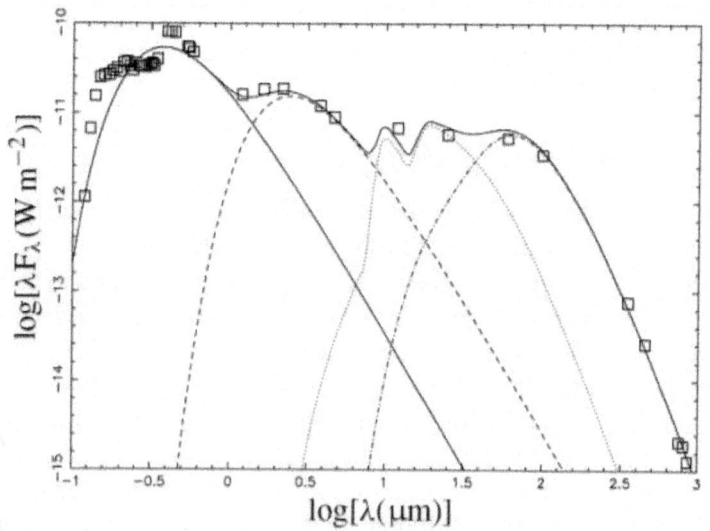

Figure. 1: Results from Dullemond et al. (2001) model applied to the star AB Aurigae. The Synthetic SED is the sum of its components: star emission (continuous thin line);

rim emission (dashed line); disc cold layer emission (dot–dashed line); and the disc hot layer emission (dotted line). The observational data in various wavelengths is represented by squares (Hetem & Gregorio-Hetem 2007).

IMPLEMENTATION

The GA code was designed and built to find the best disk parameters, namely $S = \{R_D; M_D; \theta;\ p;\ T_{rim},\ d;\ M_\star;\ L_\star;\ T_\star\}$, as discussed in subsection 2.1. However, some of these parameters are already known: the stellar parameters $d,\ L_\star;$ and $T_\star$ are adopted from observations and easily found in literature. Essentially, the GA method used implements a ^{2} minimization of the SED fitting provided by the Dulle mond et al. (2001) model. The main structures used to manipulate the data are linked lists containing the solutions (parameter set, adaptation level, χ^2_i, and the genetic operator, Φ_i), expressed by

$$M_i = \left\{ \left(R_{Di}, \theta_i, M_{Di}, p_i, T_i \right), \left(\chi_i^2, \Phi_i \right) \right\}$$

(1)

where F_j is the observed flux at wavelength λ_j, N is the number of observed data points, and φ_{ij} is the calculated flux for the solution S_j. The smallest S_i is assumed to be the gof, the goodness-of-fit measure for that generation. The evaluation function is applied to all individuals, and then the judgement procedure sorts the list by increasing X^2. It also sets one

of the genetic operators to the field Φ_i: copy, crossover, mutation or termination. Each Φ is attributed to a fraction of the number of individuals following the values suggested by Koza (1994), Bentley & Corne (2002) and references therein. With the genetic operators chosen, the next generation is evaluated by applying specific rules according to the genetic operators. The copy operator uses an elitist selection, as the solutions with the smallest x^2i are copied to the next generation. For the crossover operator, a random mix of two distinct individuals' genes is built. The mutation operator copies the original individual, except for one of the genes, which is randomly changed. The process loop continues to build new generations until the end condition is reached, as illustrated by the schematic view in Figureure. 2.

We also can estimate the error bars in the final results by analysing the x^2 behaviour as a function of the parameter variation. Then one can determine the confidence levels of a given parameter, as suggested by Press et al. (1995). Once the GA end condition has been reached, one can evaluate the inverse of the Hessian matrix $[C] \equiv [\alpha]^{-1}$ whose components are given by

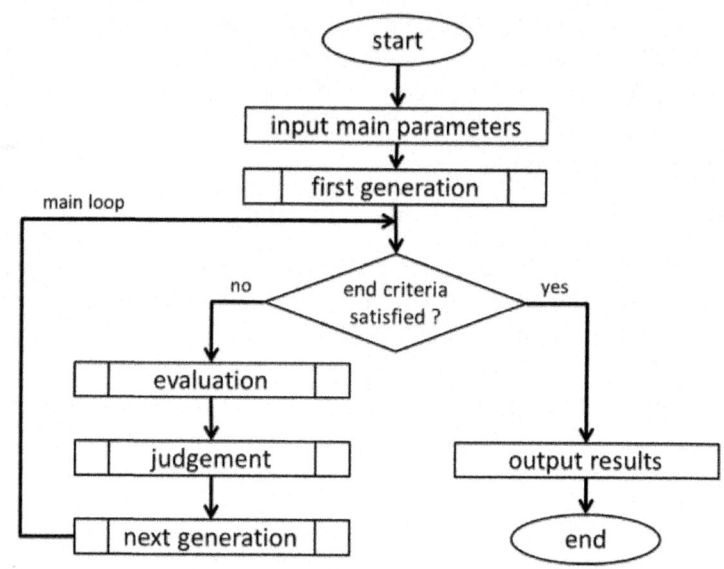

Figure. 2: Main steps of a generic GA (adapted from Hetem & Gregorio-Hetem 2007).

$$\alpha_{ij} = \sum_{k=1}^{N} \left(\frac{\partial y(\lambda_k)}{\partial a_i} \frac{\partial y(\lambda_k)}{\partial a_j} \right)$$

$$(3)$$

where $\partial y(\lambda_k)/\partial a_i$ is the partial derivative of the SED with respect to parameter a_i at $\lambda = \lambda_k$, and N is the number of observed data points. The main diagonal of C can be used to estimate the error bars on each parameter by $\sigma_i \cong C^{1/2}/N$. We estimated the error bars for the 1σ confidence level and the respective disc parameters for AB Aurigae, resulting in $M_D = 0.1 \pm 0.004 M_\odot$, $R_D = 400\pm44$ AU, $\theta = 65\pm3°$, and $T_{rim} = 1500\pm26$ K, and these results are in agreement with the error-bar estimation provided by the surface contour levels described below. Figure. 3 presents the contour levels of the gof(M_D, R_D) surface calculated for a set of 400

random pairs of disc mass and radius around the parameters for the AB Aurigae model taken from Dominik et al. (2003). The result at the minimum is gof ~ 0.046, what means that the error bar estimation converged to a narrow range around the parameter set.

We also applied the described GA method to a four other stars, in order to verify the quality of the fitting for objects showing different SED shapes and different levels of infrared excess. Our set was chosen by the slope of their near-infrared SED. The infrared excess in Herbig Be stars is the result of a spherical dusty envelope (van den Ancker et al. 2001), whereas a thickedge flared disc are characteristic of Herbig Ae. With this in mind, we selected A-type or late-B-type stars from the Pico dos Dias Survey sample (Gregorio-Hetem et al. 1992; Torres et al. 1995; Torres 1998) to apply the GA SED fitting. The results are presented in table 1 together with their corresponding gofs (see Figureure 4).

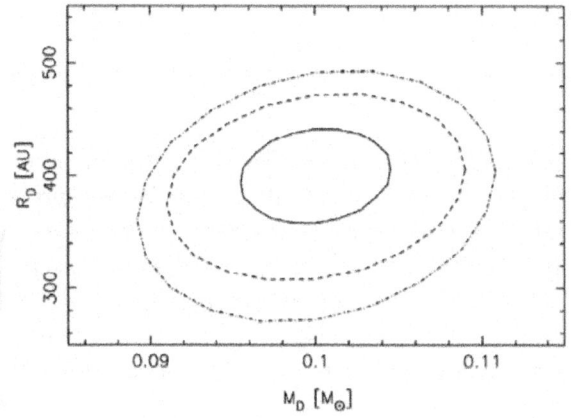

Figure. 3: Contour levels gof(M_D, R_D) estimated for AB Aurigae presenting the confidence levels $x^2(68\%)= 0.082$ (continuous line), $x^2(90\%)= 0.15$ (dashed) and $X^2(99\%)= 0.21$ (dot–dashed) (Hetem & Gregorio-Hetem 2007).

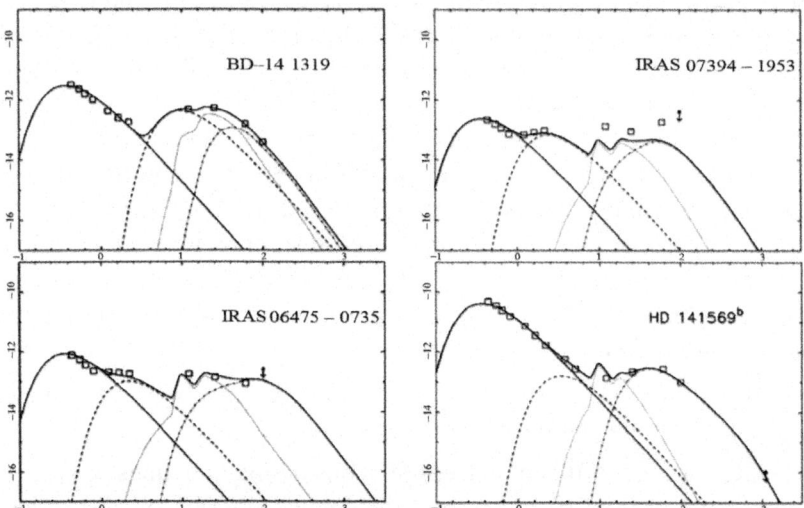

Figure. 4: GA SEDs obtained the stars BD-14 1319, IRAS 07394-1953, IRAS 06475-0735 and HD 141569. The plots are given as $log[\lambda F_\lambda (Wm^{-2})]$ versus $log[\lambda(\mu m)]$ (Hetem & Gregorio-Hetem 2007).

Table 1: Obtained parameters for the chosen stars (Hetem & Gregorio-Hetem 2007)

PDS	Name	$M_\star$ ($M_\odot$)	R_D (AU)	M_D ($M_\odot$)	T_{rim} (K)	θ (°)	p	gof
398	HD 141569	2.4	13	0.06	1085	0.6	-2.0	0.006
022	BD−14 1319	2.8	690	0.003	380	40	-10.	0.006
130	IRAS 06475−0735	2.0	309	0.20	1705	53	-1.5	0.016
257	IRAS 07394−1953	2.0	859	0.64	1838	47	-2.0	0.098

Abundances and Metallicities of young stars via Spectral Synthesis

This subsection is based on the published work The use of Genetic Algorithms and Spectral Synthesis in the Calculation of Abundances and Metallicities of T Tauri stars (Hetem & Gregorio-Hetem 2009). In the previous subsection, we presented a method that uses a calculation technique based on GA aiming to optimize the parameters estimation of protoplanetary disks of T Tauri stars. Inspired by the success of that application, which gives accurate and efficient calculations, we decided to develop a similar method to determine atomic stellar abundances.

Artificial spectra as a measurement tool

In astrophysics, the absorption spectra are obtained and employed as an analytical chemistry tool to determine the presence of atoms and ions in stellar atmospheres and, if possible, to quantify the amount of the atoms present. In stellar atmospheres, each element produces a number of spectrum absorption lines, at wavelengths which can be measured with extreme accuracy when compared to spectra emission tables provided by laboratory experiments. The presence of a given element in the star atmosphere can be verified (and measured) by looking for its absorption lines at the correct wavelength. The hydrogen is present in all stars by its Balmer absorption lines, and is often used to calibrate the measurements. An example of a high-resolution spectrum is presented in Figureure 5. The way astrophysics use to calculate the abundances of atoms in stars follows the steps:

1. Obtain the star spectrum in a given range (or ranges) of wavelength, where the lines of the elements in study should be;

2. Generate an artificial spectrum, considering the lines whose origin are the desired elements and the known physics of absorption line production;

3. Compare the artificial and observed spectra. Here a simple X^2 test is enough to compute a general comparison index;

4. Use a GA methodology to optimize the artificial spectrum in order to minimize the differences with the observed spectrum (the inverted problem, subsection 2.1);

5. Once the optimization methodology reaches its goals, consider the elemental parameters (density, temperature, ionization, etc) as the measures of the elements in the stellar atmospheres.

Inverting the problem

From our discussion on section 2, one can see that generating a synthetic spectrum is a Pproblem, as the result is obtained from a set of parameters, and no more computing is need. The generation time is obviously finite, and there are a number of very efficient software tools that do that. The only care to be taken is to assure that the artificial spectrum has the same wavelength resolution of the observed spectrum, in order to simplify the future comparison. The above mentioned step 4, a methodology to optimize the artificial spectrum, is the trick point. If one wants to use GA so solve the abundances problem, it is necessary to invert the P-problem, that is, it is necessary to use the artificial spectrum generation tool as an external routine of a bigger and more complex algorithm. The algorithm used to this task is presented in Figureure 6.

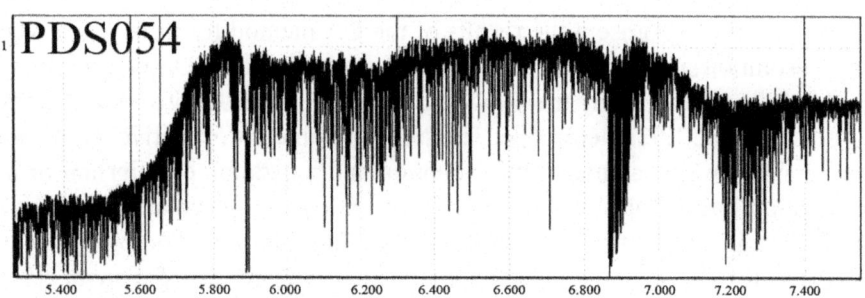

Figure. 5: FEROS spectrum for star PDS054 (Rojas et al. 2008).

Establishing the abundances of each element as the parameters to be found, one individual in the GA terminology is the set of all elemental abundances added to some atmospheric parameters. The initial parameter set is used to build the first generation with 100 individuals. The evaluator routine creates a synthetic spectrum whose entries are the genetic data in each chromosome. This task is performed by calling the elected spectral tool. There are a number of very efficient software tools that can be chosen. In our application, the abundances of chemical species are determined by using the spectral synthesis software SPECTRUM provided by Corbally (Gray & Corbally 1994) and the atmosphere model software ATLAS9 from Kurucz (1993).

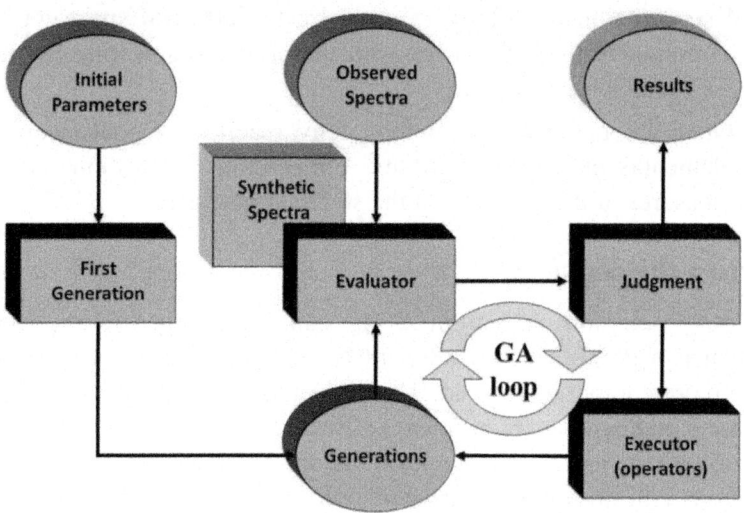

Figure. 6: Main blocks of a GA code to fit multi-band spectra of T Tauri stars (adapted from Hetem & Gregorio-Hetem 2009).

Results

In this section we present the results of the GA method for three stars, whose highresolution spectra were obtained at European Southern Observatory (ESO) in La Silla, Chile, with the Fibber Extended Range Optical Spectrograph (FEROS) at the 1.52m telescope. The stellar parameters (effective temperature and gravity) were calculated by excitation and ionization equilibrium of iron absorption lines (Rojas et al. 2008). The atomic and molecular line data were mainly from the National Institute of Standards and Technology1 and the Kurucz site[2.] The solar atomic abundances are from Grevesse & Sauval (1998), and the hyperfine structure constants were taken from Dembczyński et al. (1979) and Luc & Gerstenkorn (1972). The atmosphere models where obtained from the Kurucz library. Specific atmosphere models were calculated through a GNU-Linux porting of the ATLAS9 program (Kurucz 1993). The method performs a multi-range fitting of specific regions of the observed spectrum, looking for best fit. The demands and commands to SPECTRUM are only those for generating the specific regions of interest, but the χ^2 comparing index is evaluated over all wavelength ranges. Figureures 7 and 8 present the results for some stars on chosen lines. The metallicities and abundances found for the stars are compatible with those previously obtained for this particular sample. These preliminary results, achieved by using the GA technique, indicate the efficiency of the method. In the future, we intend to use the method in a larger sample of T Tauri stars

APPLICATIONS ON ROCKET ENGINE ENGINEERING

This section presents two solutions in applying GAs in the aerospace area, both concerning the fuel pumping in liquid propellant rocket engines. There are many choices to be done in the design of a high performance fuel pump, being one of them the type of pump. Two different types of pumps were modelled: the Harrington pumps and the turbo pumps. Both present a complex design methodology, which includes: tabled functions interpolations, numerical integrals and constructive material choices.

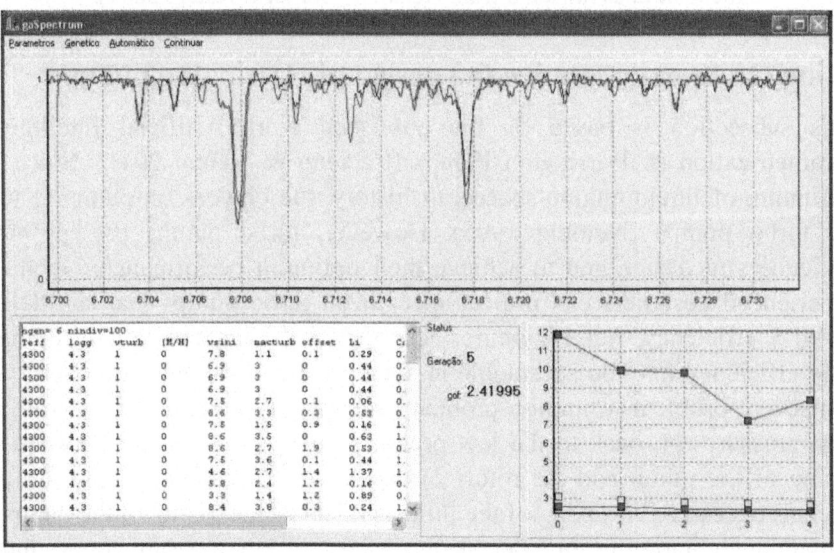

Figure. 7: Main screen of the program GASpectrum after five generations. The upper panel presents the spectra: the blue line represents the observed spectrum and the red line represents the best individual spectrum (adapted from Hetem & Gregorio-Hetem 2009).

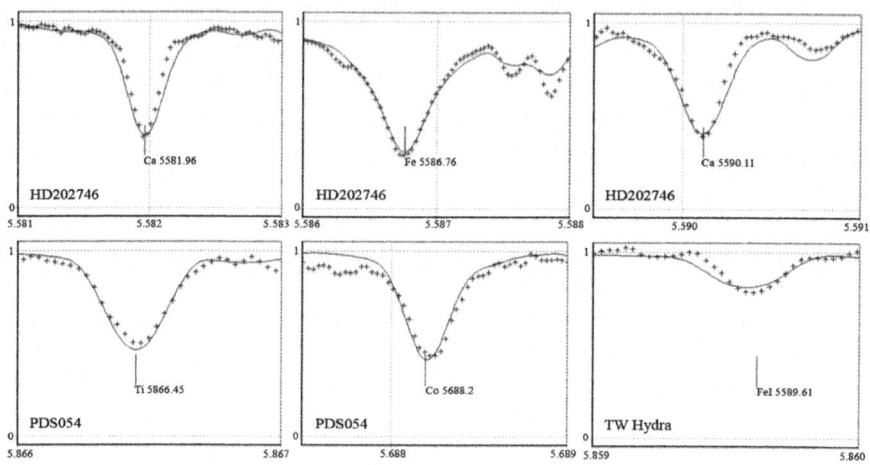

Figure. 8: Main results for stars HD202746, PDS054 and TW Hydra, on calcium, iron, titanium and cobalt lines (adapted from Hetem & Gregorio-Hetem 2009).

Using GA to parameterize the design of Harrington pumps

This subsection is based on the published work Artificial Intelligence Parametrization of Harrington Pumps (Caetano & Hetem 2011). Since the beginning of liquid engine spacecraft history, the choices on pumping were the turbo pumps (Neufeld 1995). However, turbo pumps present many difficulties to design and to achieve their optimum performance. Good and experienced designers can project specialized turbo pumps that can deliver 70-90% efficiency, but Figureures less than half that are not uncommon. Low efficiency may be acceptable in some applications, but in rocketry this is a severe problem. Common problems include: 1) excessive flow from the high pressure rim back to the low pressure inlet along the gap between the casing of the pump and the rotor; 2) excessive recirculation of the fluid at inlet; 3) excessive vortexing of the fluid as it leaves the casing of the pump; 4) damaging cavitation to impeller blade surfaces in low pressure zones; and 5) critical shaping of the rotor itself is hardly precise (see the many examples and demonstrations presented by Dixon & Hall (2010) for a better understanding of these concerns). On the other end, the options are the pressurized tanks. In this choice, the fuel and oxidizer reservoir are filled charged with a high pressure gas (helium or nitrogen) that pushes the fluid to the thrust chamber. So, it is easy to see that the tank output fuel pressure drops as the rocket engine consumes its content. As an option, the designer can increase the inside pressure, but this came also with a high cost in material (due to tank thickness) and instability. Actually, pressurized propellant tanks are used on small rockets like the last stages on space missions. As an elegant intermediate solution between these two extremes, Harrington (2003) presented a design fills the gap between the pressure fed and the turbo pumps. This solution also has the advantage of lowering the costs of a rocket project, keeping low weight and without the high complexity of a turbo pump, whose operation, theoretical concerns and constructive details are explained in next section.

Using GA to parameterize the design of Harrington pumps

This subsection is based on the published work Artificial Intelligence Parametrization of Harrington Pumps (Caetano & Hetem 2011). Since the beginning of liquid engine spacecraft history, the choices on pumping were the turbo pumps (Neufeld 1995). However, turbo pumps present many difficulties to design and to achieve their optimum performance. Good and experienced designers can project specialized turbo pumps that can deliver 70-90% efficiency, but Figureures less than half that are not uncommon. Low efficiency may be acceptable in some applications, but in rocketry this is a severe problem. Common problems include: 1) excessive flow from the high pressure rim back to the low pressure inlet along the gap between the casing of the pump and the rotor; 2) excessive recirculation of the fluid at inlet; 3) excessive vortexing of the fluid as it leaves the casing of the pump; 4) damaging cavitation to impeller blade surfaces in low pressure zones; and 5) critical shaping of the rotor itself is hardly precise (see the many examples and demonstrations presented by Dixon & Hall (2010) for a better understanding of these concerns). On the other end, the options are the pressurized tanks. In this choice, the fuel and oxidizer reservoir are filled charged with a high pressure gas (helium or nitrogen) that pushes the fluid to the thrust chamber. So, it is easy to see that the tank output fuel pressure drops as the rocket engine consumes its content. As an option, the designer can increase the inside pressure, but this came also with a high cost in material (due to tank thickness) and instability. Actually, pressurized propellant tanks are used on small rockets like the last stages on space missions. As an elegant intermediate solution between these two extremes, Harrington (2003) presented a design fills the gap between the pressure fed and the turbo pumps. This solution also has the advantage of lowering the costs of a rocket project, keeping low weight and without the high complexity of a turbo pump, whose operation, theoretical concerns and constructive details are explained in next section.

The model: Pump constructive details

Designing a Harrington pump is simple, but the optimization process is not (as expected: a P-problem and a NP-problem respectively). A pump with a small chamber must be filled and vented quickly, with minimal head loss through the gas and liquid valves and plumbing. Making the pump cycle as fast as possible would make it lightweight, but higher flow velocities cause problems (Harrington 2003).

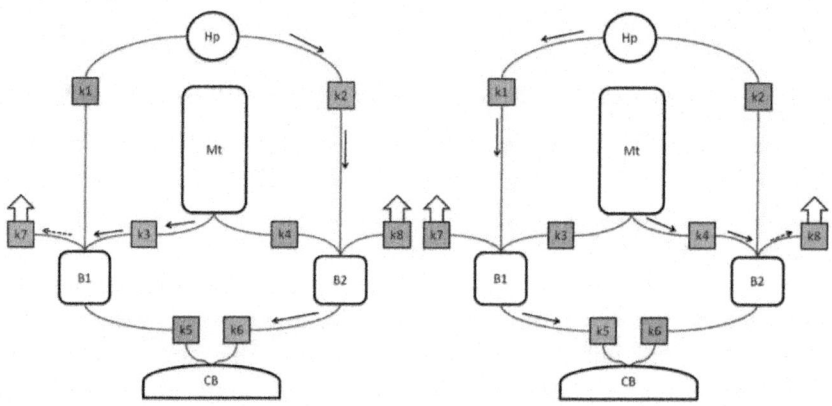

Figure. 9: Schematic view and operation of a Harrington pump, with its chambers (B1 and B2) and valves (k1-8). The main rocket fuel tank is represented by Mt whereas Hp represents a high pressure gas generator. The two states are presented. Left: B1 is being filled by Mt while B2 is feeding the combustion chamber. Right: B1 is feeding the combustion chamber while B2 is being filled by Mt. the arrows indicate the flow. (Caetano & Hetem 2011).

Table 2.: Derived model parameters for the sample (Caetano & Hetem 2011)

valve	state 1	state 2
k1	closed	open
k2	open	closed
k3	open	closed
k4	closed	open
k5	closed	open
k6	open	closed
k7	open	closed
k8	closed	open

The choice of pump tanks material plays an important role, as its mass density and stress coefficients are the main keys in the pump design. The main tank pressure (about 300 kPa) and the area of the inlet valves set up the limits for the maximum inflow rate. If the inflow velocity is increased this can cause the propellant to be aerated, what is not desirable for the proper working of the engine. The extra volume of pressurized gas in the pump chamber should be small to minimize gas usage, but if it is too small, there will be a loss of propellant through the vent. The primary parameters for the calculations are the state changing cycle t_{cy}, the volume flow determined by the rocket engine needs, Q, the specific impulse of the propellants, I_{sp}, at the fuel pressure, Pf, the fuel mass density, ρ_f, the thrust, T, and the material properties: the mass density, ρ_c, and stress coefficient, σ_c. From these parameters, considering the pump chambers are spherical, one can instantly obtain the diameter of one chamber:

$$D_c = \sqrt[3]{6\frac{\int_0^{t_{cy}} Q dt}{\pi}},$$

(4)

where the integral results in the chamber volume, and for the simplest case of steady flow, it resumes to $V_c = Q.t_{cy}$. Knowing the diameter and applying the stress formulae from Young (1989), the chamber walls thickness can be obtained by

$$t_w = \frac{P_f D_c}{\sigma_c},$$

(5)

and the total chamber mass by

$$M_c = \pi t_w D_c^2 \rho_c.$$

(6)

To obtain the thrust, one can apply the momentum equation for the case of ideal expansion, and:

$$T = gQI_{sp}\rho_f,$$

(7)

where g represents the gravity acceleration. Manipulation of these expressions and an estimative of the relative weight of the valves and other accessories lead to expression 7 from Harrington (2003), the pump thrust to weight ratio:

$$\frac{T}{W} = 0.43 \frac{g l_{sp} \rho_f \, \sigma_c}{P_f T_{cy} \, \rho_c},$$

$$(8)$$

that is to be optimized. The total pump mass is Mp=1.56 Mc, and the mass flow can be easily obtained by m Q= ρ_f & . The expressions (4)-(8) were coded in a program to test the feasibility of this set of equations as a model. Table 3 presents the results obtained for typical parameter values. These results are in agreement with rocket engine pump literature (Griffinand & French 1991; Sutton 1986).

Table 3: Test values for the pump model and results.

Entry parameters			Model results		
t_{cy}	5	s	V_c	0,016667	m³
Q	200	l/min	D_c	31,69203	cm
I_{sp}	285	s	t_w	0,090549	cm
ρ_f[1]	935	kg/m³	M_c	0,8	kg
P_f	4	Mpa	M_p	1,248	kg
T	8800	N	T	8704,85	N
σ_c[2]	350	MPa	T/W	8718,8	
ρ_c[2]	2,8	g/cm³	$\dot{m}$	3,116667	kg/s

[1] Propellant mixture: LOX/RP-1

[2] 2219 Aluminum alloy

GA optimization method Here we describe de Genetic Algorithm (GA) optimization method and the formalism applied to code the problem to its needs. The pump parameters we want to find are a subset of those described as primary parameters: the state changing cycle, t_{cy}, the fuel pressure, Pf, the fuel mass density, ρ_f and the material properties: the mass density, ρc, and stress coefficient, σ_c. These are the GA free parameters, formally

$$\Lambda = \{t_{cy}, P_f, \rho_c, \sigma_c\},$$

$$(9)$$

known as the parameter set. The technique used to work with the material parameters, ρ_c and σ_c, are explained in sub-section 4.1.4. The obtained pump must deliver a desired mass rate, , of a given propellant, ρ_f, and must be made of a given material, ρ_c and σ_c. Some variables are project dependent, like the volume flow, Q, the specific impulse of the propellants, I_{sp}, at the fuel pressure, and the thrust, T. These three parameters are those the rocket engine designer should define to specify the pump he needs. Differently from the first parameters described on the above paragraph, these values cannot be altered by the algorithm, and can be included in another group, the constant set:

$$\Psi = \{Q, I_{sp}, T\} .$$

Another group of variables is need: the result set. These are the values that are obtained by running the model code:

$$\Gamma = \{V_c, D_c, t_w, M_c, M_p, T, T / W, \dot{m}\} . \tag{11}$$

To satisfy the GA formalism, one must write down the model that describes the necessary transformations to obtain Γ from Ψ and Λ, or $\Lambda = f(\Psi, \Gamma)$.

Now we explain how the GA method was implemented in the Harrington pump model described above. We first clarify the GA nomenclature in the field of pump design. A parameter (e.g. volume flow) corresponds to the concept of a 'gene', and a change in a parameter is a 'mutation'. A parameter set that yields a possible solution corresponds to a 'chromosome', our Λ. An 'individual' is a solution that is composed of one parameter set and two additional GA control variables. One of these variables is χ^2, which refers to the 'adaptation' level. The other control variable is Φ, the genetic operator. The term 'generation' means 'all the individuals' (or all the solutions) present in a given iteration.

The code uses the parameters described in (9), namely $\Lambda = \{t_{cy}, P_f, \rho_c, \sigma_c\}$ Essentially, the GA method presented herein implements a X^2 minimization of the comparison between the desired results $\Gamma_0 = \{V_c, D_c, t_w, M_c, M_p, T, T / W, \dot{m}\}$, and the results obtained by the application of expressions (4) to (8), the model results. There are three main advantages of using a GA for this task: (i) the GA method potentially browses the whole permitted parameter space, better avoiding the 'traps' of local minima; (ii) the method is not affected by changes in the model; (iii) the GA implementation does not need to compute the derivatives of χ^2 (such as $\partial \chi^2 / \partial P_f$, for example) required by the usual methods. This fact simplifies the code and minimizes computer errors caused by gradient calculations.

The main structures used to manipulate the data are linked lists containing the solutions (parameter set, adaptation level, χ_i^2 and the genetic operator, Φ_i, expressed by $S_i = \{\Psi, \Lambda_i, \Gamma_i, (\chi_i^2, \Phi_i)\}$ where S_i denotes the ith solution. Following Goldberg (1989) and Hetem & Gregorio-Hetem (2007), the code starts with the construction of the first generation, where all parameters are randomly chosen within an allowed range (for example, 15 cm $< D_c < 30$ cm).

Here, the number of parameter sets in the first generation is assumed to be 100. In the next step, the evaluation function runs the model for each solution, and compares the synthetic Γ_0, to find χ^2, using a modified expression given by Press et al. (1995):

$$\chi_i^2 = \frac{1}{n_p} \sum_{j=1}^{n_p} \left(\frac{\Gamma_{0j} - \Gamma_{ij}}{\Gamma_{0j}} \right)^2 ,$$

(12)

where n_p is the number of values in the result set, Γ_{0j}, is the desired value on position j (e.g. $\Gamma_{01} = V_c$), and Γ_{ij} is the calculated value for the solution S_i. The smallest χ^2 corresponds to the goodness-of-fit, or simply gof. The gof values express how each individual is adapted, or how close each solution is, to the best solution (Bentley & Corne 2002). For the value of T/W, which we want to optimize, it is enough to establish a corresponding to Γ_{0j} very high.

A judgment function then determines the genetic operator Φ to be applied to a solution. Its values can be 'copy': the individual remains the same in the next generation; 'crossover': the individual is elected to change a number of genes (parameters) with another individual, creating a new one; 'mutation': one of its genes is randomly changed; or 'termination': none of the genes continue to subsequent generations. The chosen action is expressed by the Φ_i variable, associated with each individual. The next step is to evolve the current generation (k) to the next (k + 1) one, which is done through a multi-dimensional function β that considers the solutions and the genetic operators. Formally,

$$[S_1, S_2, \ldots, S_N]_{k+1} = \beta \left[(S_1, \Phi_1), (S_2, \Phi_2), \ldots, (S_N, \Phi_N) \right]_k .$$

(13)

As soon as a new generation is ready, the evaluation function is reapplied, and the algorithm repeats the described actions until an end-of-loop condition is reached. The end condition can be based on the number of iterations or the quality (a low level for the χ_i^2 values).

The choice of chamber constructive material

The main material properties, the mass density, ρ_c, and stress coefficient, σ_c, can also be chosen by the GA. Instead of working directly with these parameters, it was created a material parameter, K_c, an integer that points to a density-stress database. So, our new parameter set becomes

$$\Lambda = \{t_{cy}, P_f, \rho_c(K_c), \sigma_c(K_c)\},$$

(14)

or simply

$$\Lambda = \{t_{cy}, P_f, K_c\}.$$

(15)

As K^c is a discrete value, it was needed to build special routines to manipulate the genes in the first generation and in mutation events.

RESULTS AND CONCLUSION

Table 4 presents the main results for a GA run of 20 generations. The values are in agreement with the expected for the pump. The material chosen for the chambers was cooper 99.9%. A typical running with about 100 generation is achieved in ~5 seconds in a simple laptop computer.

Table 4: GA result values for the pump model.

t_{cy}	8.2	s	Vc	0.00393786	m³
Q	200	1/min	Dc	0.195924	cm
I_{sp}	285	s	tw	0,089	cm
ρ_f [1]	935	kg/m³	Mc	0.957973	kg
P_f	4	Mpa	Mp	1.49444	kg
σ_c [2]	350	MPa	T/W	843.227	
ρ_c [2]	2,8	g/cm³	$\dot{m}$	0.448098	kg/s

[1] Propellant mixture: LOX/RP-1

[2] Copper 99.9% Cu

The GA proved to be efficient, and due to the method itself being independent of model complexity, it certainly can be used in future implementations of pump design. Future evolutions and increasing complexity of the model, like thermal transfer and realistic valves, can benefit of GA robustness and reliability. The next step in this work is to enhance the model with more realistic and specific trends. It is expected to incorporate non-linear functions, differential equations and integrals. Also tabled functions are not far from what can be found in a pump project, with its intrinsic interpolations. The overall problem of finding parameters for a pump design can easily turn to a NP-Problem, that is a problem that is very difficult to find a solution, but, once one has a candidate to solution it is easy to verify if it is a good solution.

Using GA to parameterize the design of turbo pumps to be used in rocket engines This subsection is based on the published work Parametric Design of Rocket Engine Turbo pumps with Genetic Algorithms (Burian et al. 2011). Turbo pumping in high-thrust, long-duration liquid propellant rocket engine applications, generally results in lower weights and higher performance when compared to pressurized gas feed systems. Turbo pump feed systems require only relatively low pump-inlet pressures, and thus propellant-tank pressures, while the major portion of the pressure required at the thrust chamber inlets is supplied by the pumps, saving considerable vehicle weight. As stated by Huzel & Huang (1967) the best performing turbo pump system is defined as that which affords the heaviest payload for a vehicle with a given thrust level, range or velocity increment: gross stage take-off weight; and thrust chamber specific impulse (based on propellant combination, mixture ratio, and chamber operating efficiency). The particular arrangement or geometry of the major turbo pump components is related to their selection process (Logan & Roy 2003). Some complex designs, like the SSME-Space Shuttle Main Engine, have a multiple stage pump, but most propellant pumps have a singlestage main impeller. Eventually, one or more design limits are reached which requires more iteration, each with a new changed parameter or approach. For a better example, see table 5 which presents some data from the V2 (II world war German missile) alcohol pump

Table 5. Parameters from the alcohol V2 pump, adapted from Sutton & Biblarz (2001).

Parameter	value
impeller diameter	34 cm
rotation	5000 rpm
performance	265 kW
delivery	50 kg/s
delivery pressure	25 atm

This subsection considers the development of a software tool based on GA to assist the determination of the excellent parameters for the conFigureuration of turbo pumps in engines for liquid propellant rockets. We present the first version, which considers the calculation of the main parameters of a compressor stage

The model

The pump compressor model used in this work is based on chapter 10 of Sutton & Biblarz (2001). This model provides a coherent basis for the modeling, and is sufficiently complex to be used as a valid test on the further parameter optimizing step. The pump parameters we want to find are: the inlet compressor diameter, d_1, the compressor outlet diameter, d_2, the fluid input velocity, v_1, the suction specific speed, S, the shaft cross section, AS_1, the pressure in the main tank, P_t, the total fluid friction (viscosity included) due to flow through the pipes, valves, etc, P_f, the pressure due to the tank elevation from the pump inlet, P_e. In particular, this last parameter leads to project insights concerning the pump position inside the rocket. These are the GA free parameters, formally $\Lambda = \{d_1, d_2, v_1, S, dS_1, P_t, P_f, P_e\}$, known as the parameter set. The obtained compressor must deliver a desired mass rate, m& , and, from an input pressure P_1, generate a flow with an output pressure P_2. Some constants shall be considered, like the fluid mass density, ρ, and the external gravity, g_0. We assumed as fluid the ethanol (C_2H_6OH) due to its green properties and green results. These three parameters are those the rocket engine designer should define to specify the compressor he needs. Differently from the first eight parameters described on the above paragraph, these values cannot be altered by the algorithm, and can be included in another group, the result set $\Gamma = \{\dot{m}, P_1, P_2\}$. To satisfy the GA formalism, one must write down the model, or the formalism that describes the necessary transformations to obtain Γ from Λ, or $\Gamma = f(\Lambda)$. One can obtain these expressions following Sutton & Biblarz (2001) model and converting their expressions. First, the pressures should be converted to heads, or the height necessary to the fluid to cause a given pressure, so we define H_t, He and H_f, the tank head, the elevation head and the friction head, respectively, that can be obtained by

$$P_t = \int_{H_t} g_0 \rho dh ,$$

(16)

$$P_e = \int_{H_e} g_0 \rho dh ,$$

(17)

And

$$P_f = \int_{H_f} g_0 \rho dh .$$

(18)

The effective area of the inlet is given by

$$A_{1eff} = \frac{1}{4}\frac{d_1^2}{\pi} - A_{S1},$$

(19)

which determines the volume flow

$$Q = \iint\limits_{A_{1eff}} v_1 dA .$$

(20)

Then, the absolute positive head can be obtained by

$$H_1 = H_t + H_e - H_f$$

(21)

and the net positive suction head or available suction head above vapor pressure can be obtained by

$$H_s = H_t + H_e - H_f - H_v ,$$

(22)

where $\phi=3/4$ and $u_{SI}=17.827459$ are constants. u_{SI} is necessary due to SI convertions (see Sutton & Biblarz 2001, eq. 10-7). This last expression allows us to obtain N_{rad}/s, the shaft speed in radians per second. The impeller vane tip speed is given by

$$u = \frac{1}{2}d_2 N_{rad/s} .$$

(23)

With u, we can evaluate the head delivered by the pump

$$\Delta H = \frac{u^2}{\psi g_0} ,$$

(24)

where ψ has values between 0.90 and 1.10 for different designs. As for many pumps, $\psi = 1.0$, we adopt this value.

At this point, we are able to obtain all the final results, $\Gamma = \{\dot{m}, P_1, P_2\}$:

$$P_1 = H_1 g_0 \rho ,$$

(25)

$$P_2 = (\Delta H + H_1) g_0 \rho ,$$

(26)

And

$$\dot{m} = \rho Q.$$

(27)

It is also interesting to evaluate the shaft specific speed

$$N_s = \frac{u_{SI}\sqrt{Q}}{H_{SR}^{\phi}},$$

(28)

which, with the aid of table 10⁻² of Sutton & Biblarz (2001), defines the pump and impeller type.

Results and conclusion

We built a computer code to optimize equations in the same way it was done to the Harrington pumps (see subsection 4.1). The resulting parameters obtained from the GA code where in good agreement with what is expected for this kind of project. Some comparisons between GA results and correct results are presented in table 6.

Table 6: Comparison between obtained results (GA) and correct answer (Γ0) for an ethanol compressor.

		$\dot{m}$ (kg/s)	P_1 (Pa)	P_2 (Pa)	mean error (%)
Correct answer		226,8	342669	6816870	
generations	10	228,1	342345	6816450	0,22
	20	227,5	342360	6816440	0,13
	50	227,1	342601	6816890	0,05
	100	226,9	342670	6816880	0,01

Evidently, for the simple definitions presented for this model, one does not need a sophisticated method as described to obtain a good result. But, as all designers know very well, there are no simple projects, especially concerning rocket engine pumps. The next step in this work is to enhance the model with more realistic and specific trends. It is expected to incorporate non-linear functions, differential equations and integrals. Also tabled functions are not far from what can be found in a pump project, with its intrinsic interpolations. The overall problem of finding parameters for a pump design can easily turn to a NP-Problem, that is a problem that is very difficult to find a solution, but, once one has a candidate to solution it is easy to verify if it is a good solution. Again, the GA proved to be efficient, and due to the method itself being independent of model complexity, it certainly can be used in future implementations. Future evolutions and increasing complexity of the model can benefit of GA robustness and reliability.

APPLICATIONS ON ENERGY DISTRIBUTION

The application described in this section solves the problem of allocation of protective devices in electric power distribution plants. For a given power plant distribution, it is necessary to choose in which points one must place equipment for the net protection, or not.

This problem is entirely based on discrete elements – there are no floating point parameters. So, the main discussion here is how to build a chromosome syntax that can be used under the GA rules, and still be meaningful for the model. Besides, as the problem is fully discretized, there are high probabilities of finding different solutions that are equally evaluated in their adaptation function. This leads to new enhancements in the model to better evaluate the solutions, enhancing the separation between different individuals.

Using GA in the allocation of electric power protective devices

This subsection is based on the published work Automatic Allocation of Electric Power Distribution Protective Devices (Burian et al. 2010). The measurement of how well the electric power distribution system can provide a secure and adequate supply of power to satisfy the customer's requirements is called "reliability". Regarding electric power distribution systems, the electric utilities companies are responsible for the most reliable service as possible, reflecting the most advanced state of technology with reasonable cost to the end product that is the electric power3. Most utilities record outage information such as the number of outages, elapse time, and the number of customers interrupted. These data and statistics may be reported for each circuit or operating division, for comparison purposes, using the standard performance indices. The performance indices provide historical datum which can be used to determine increasing or decreasing trends and to measure whether system improvement plans have yielded expected results. The quality model we consider in this subsection uses the following indices, based on the sustained outage data: the SAIDI and SAIFI indexes, explained as follows:

SAIDI (System Average Interruption Duration Index): defined by the rate of average interruption duration per customer served per year. This index is commonly referred to as minutes of interruption per customer.

$$SAIDI = \frac{\text{Sum of Customer Interruption Durations}}{\text{Total Number of Customers Served}}$$

(29)

SAIFI (System Average Interruption Frequency Index): that defined by the rate of average number of times that a customer's service is interrupted during a reporting period per customer served in a given period (usually one year). A customer interruption is defined as one sustained interruption to one customer:

$$SAIFI = \frac{\text{Total Number of Customer Interruptions}}{\text{Total Number of Customers Served}}$$

(30)

It is easy to see that what is desired is a circuit with minimal SAIDI and SAIFI with the smaller cost in protective installed devices. The resulting circuit with these characteristics will the optimized circuit.

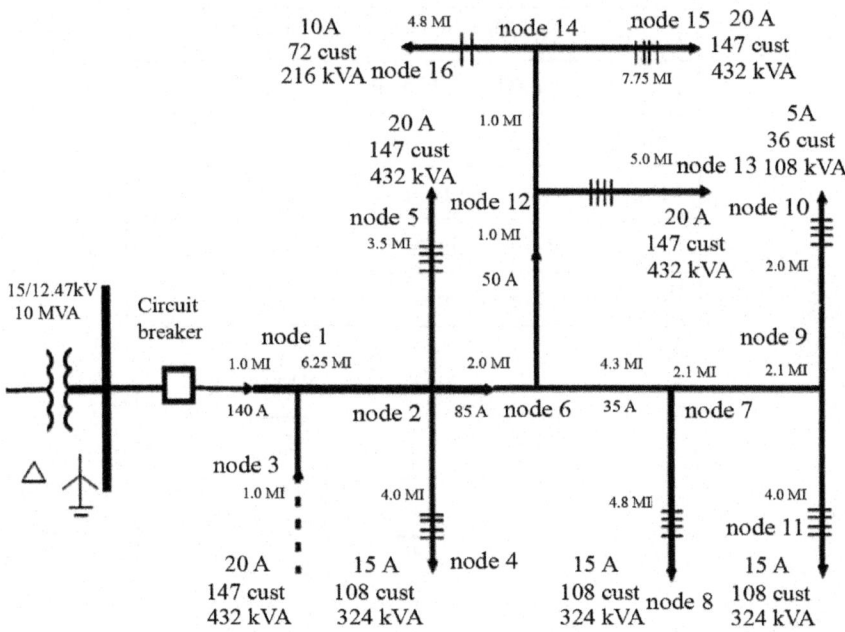

Figure. 10: Circuit with Circuit Breaker in the Electric Power Substation without Reclosing Capability, based on Bishop (1997).

The Model

The chosen model was based in the work developed by Bishop (1997) whose circuit has multiple laterals with customer's numbers and load KVA values seen on the Figureure 10. To perform the analysis one needs some statistics, like:

number of customers; placement of protective devices on the electric power utility; good possibilities to implement protective devices; distribution circuit response to the quality indices; and traditional values of repair and recover in accordance with Bishop's indices. The initial circuit used to the analysis is presented by Figureure 10, where it was considered the values of Bishop (1997) to the indices in circuits of electric power distribution with similar features in North American solutions. The used general statistical parameters are presented in table 7. As a base case analysis, the system was modelled with no reclosing of substation device. This is intended only to yield values for relative comparison with other circuits, with protective devices like recloses and fuses placed on the circuit, achieving the comparison landscape with the SAIDI and SAIFI indices.

Table 7. General statistical parameters used in the model.

Faults per circuit mile per year	0.22
Percent of permanent faults	20%
Percent of temporary faults	80%
Manual restoration time	2.0 hours
Repair time for 30 lines	3.0 hours
Repair time for 10 lines	2.5 hours

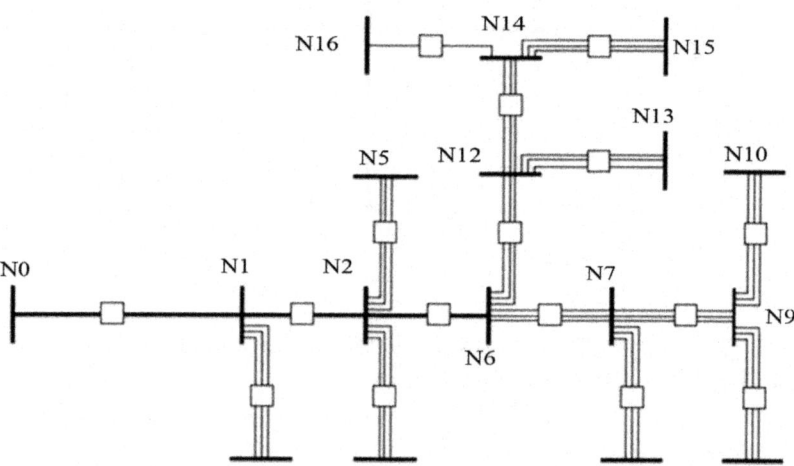

Figure. 11: Representation of the circuit of Figureure 10 with the nodes with all the possible locations for protective devices (adapted from Burian et al. 2010).

The methodology: Converting to a GA application

The first step is to provide formalism in such a way that the protective devices net could be represented by a set of genes in a chromosome Λ, and that the Bishop (1997) model could be expressed as a P-problem whose parameters are given by Λ. The solution chosen was to code the circuit as a series of nodes, designed by Ni, with i being an integer number, and to build a list of links between the nodes (see Figureure 11). The special node N0 is the main protective switch in the substation (which is present in all solutions). Each link between nodes can have a protective device, and its location is designed as $P_{i,j}$, with i and j being the two nodes that define the link. Special data structure is provided to the nodes to storage information about the number of phases, number of consumers, distance to neighbours nodes, etc. The adopted solution considers S as a ordered list of tokens, and the position in the ordered list corresponds to a location as $P_{i,j}$. Then, for the circuit of Figureure 11, one has

$$\Lambda = \left\{ \begin{matrix} P_{0,1}, P_{1,2}, P_{1,3}, P_{2,4}, P_{2,5}, P_{2,6}, P_{6,7}, P_{6,12}, P_{7,8}, \\ P_{7,9}, P_{9,10}, P_{9,11}, P_{12,13}, P_{12,14}, P_{14,15}, P_{14,16} \end{matrix} \right\}.$$

(31)

So, Λ is a finite set of tokens, and its number of elements is much smaller than the number of nodes squared[4], that assumes the role of parameter set in the P-problem. These tokens can represent a protective device to be placed in its respective circuit position. The possible devices are: main substation switch, only possible in location $P_{0,1}$ (S); fuse (F), automatic reclose switch (R) and nothing (no device).

The kind of device defines the algorithm to be used to obtain the overall cost of protective devices, and the SAIDI and SAIFI indexes according to Bishop (1997). So, each set Λ_i represents a different circuit, and applying the Bishop's algorithms one obtains a result set

$$\Gamma_i(\Lambda_i) = \{SAIDI_i, SAIFI_i, c_{Si}, c_{Ri}, c_{fi}\}.$$

(32)

where c_S, c_R and c_f are the costs of the main switch reclose switch and fuses, which are expressed in monetary "units", being one unit the cost of the a monophasic fuse. As the set Γ_i itself cannot express the degree of adaptation the individual Λ_i to the problem we want to solve, we must provide an expression to summarize Γ_i in a more convenient, single valued variable, like the gof value, described in subsection 3.1. The definition of this gof should have a monotonic behaviour as the costs and the SAIDI and SAIFI index increase. We adopted the simple expression

$$gof = \kappa_a (\text{SAIDI} + \text{SAIFI}) + \kappa_b \left(c_S + c_R + c_f \right).$$

(33)

where κ_a and κ_b are constant scale converters. Then, one can say that optimized circuit will be that one that offers the smaller gof. With this, our inverted NP-problem can be solved by looking for the individual Λ_i that presents the smaller gof. As all the parameters are limited range integer numbers (tokens), some special care must be taken in the GA routines that deal with new individuals and mutation. So, these routines where rebuild taking into account the discrete character of the chromosomes. The overall behaviour of the GA optimization code follows the algorithm proposed in Figureure 2.

Results and conclusion

The resulting optimized circuit is shown in Figureure 12, and its corresponding indexes are presented in table 8. The GA code performed the ranging of large number of solutions and conFigureurations, within the universe of about 50 generations of conFigureurations. This demonstrates the GA potential in this kind of analysis and application to discrete allocation equipment's. GA optimization techniques has been showed to be an effective technique to optimize the allocation of protective devices inside the electrical distribution systems.

Table 8: Indexes values for optimized circuit

Index	value
SAIDI	2.7694
SAIFI	1.04385
Cost S	60 units
Number S	1
Cost R	280 units
Number R	3
Cost F	25 units
Number F	9
Total Cost	365 units

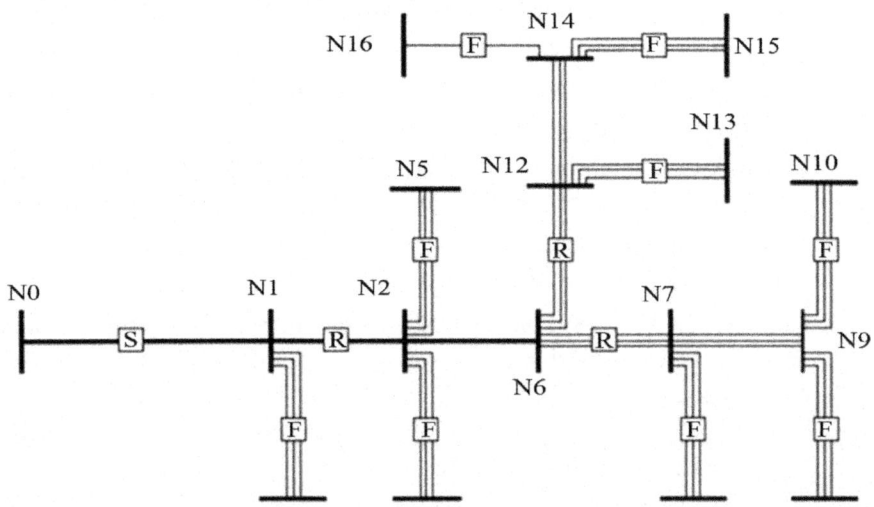

Figure. 12: Optimized circuit obtained with the GA method (adapted from Burian et al. 2010).

ACKNOWLEDGMENTS

The author wants to thank UFABC/CECS - Engineering, Modeling and Social Sciences Center of Federal University do ABC; AEB – Brazilian Space Agency / UNIESPAÇO Program; FAPESP and CNPq.

REFERENCES

1.   André P., Ward-Thompson D., Barsony M., (1993), ApJ, 406, 122 Bentley, P.J., & Corne D.W. (2002) Creative Evolutionary Systems. Morgan-Kaufmann, San Francisco.

2.   Bishop, M.T. (March 1997) Establishing Realistic Reliability Goals. The Tech Advantage 97 Conference & Electric Expo

3.   Burian, R.; Hetem, A., Caetano, C. A. C. Automatic Allocation of Electric Power Distribution Protective Devices (2010) Opatija. 33rd International Convention on Information and Communication Technology, Electronics and Microelectronics. Opatija / Abbazia : IEEE, 2010. v. 1. p. 22-22.

4.   Caetano, C.A.C., & Hetem, A. (2011) Artificial Intelligence Parametrization of Harrington Pumps, to be submitted to International Journal of Heat and Fluid (in preparation).

5.   Cook, Stephen (1971) The complexity of theorem proving procedures. Proceedings of the Third Annual ACM Symposium on Theory of

Computing. pp. 151–158.Dembczyński, J., Ertmer, W., Johann, U., Penselin, S., & Stinner, P. 1979, Z. Phys. A, 291, 207

6. Dixon, S. L.,Hall,C. A. (2010) Fluid mechanics and thermodynamics of turbomachinery 6th ed. ISBN 978-1-85617-793-1

7. Dominik C., Dullemond C. P., Waters L. B. F. M., Walch S. (2003) A&A, 398, 607 Dullemond C. P., Dominik C., Natta A. (2001) ApJ, 560, 957

8. Garey, M. R., & Johnson, D. S. (1979) Computers and Intractability: A Guide to the Theory of NPcompleteness. W. H. Freeman.

9. Goldberg D. E. (1989) Genetic Algorithms in Search, Optimization and Machine Learning. Addison-Wesley Longman, Boston, MA Gray, R.O., & Corbally, C.J. (1994) AJ, 107, 742.

10. Gregorio-Hetem J., Lépine J. R. D., Quast G. R., Torres C. A. O., de la Reza R. (1992) AJ, 103, 549

11. Grevesse, N. & Sauval, A.J. (1998) Space Science Reviews 85, 161 Griffinand, M.D., & French, J.R. (1991) Space Vehicle Design, AIAA. Harrington, S. (2003) Pistonless Dual Chamber Rocket Fuel Pump, 39th AIAA/ASME/SAE/ASEE Hoint Propulsion Conference and Exhibit. AIAA 2003- 4479.

12. Hetem, A., & Gregorio-Hetem, J. (2007) The use of genetic algorithms to model protoplanetary discs, MNRAS 382, 1707–1718 (2007) doi:10.1111/j.1365- 2966.2007.12442.x

13. Hetem Jr, A. ; Gregorio-Hetem, J. (2009) The use of Genetic Algorithms and Spectral Synthesis in the Calculantion of Abundances and Metallicities of T Tauri stars. In: Young stars, Brown Dwarfs and Protoplanetary Disks Special Session 7 - IAU XXVII General Assembly, 2009, Rio de Janeiro - RJ. IAU XXVII General Assembly Abstract Book. Paris - France : International Astronomical Union, 2009. v. 1. p. 481-481.

14. Huzel, D.K. & Huang, D.H. (1967) Design of Liquid Propellant Rocket Engines, Rocketdyne Division, North American Aviation, Inc

15. Koza J. R. (1994) Genetic Programming II: Automatic Discovery of Reusable Programs. MIT Press. Kurucz, R. L. (1993) CD-ROM 13, Atlas9 Stellar Atmosphere Programs and 2 km/s Grid (Cambridge: Smithsonian Astrophys. Obs.) Lada C. J., Wilking B. A. (1984) ApJ, 287, 610

16. Logan, E., Jr., & Roy, R, (eds) (2003) Handbook of Turbomachinery (Second Edition Revised and Expanded), Marcel Dekker, Inc.

17. Luc, P. & Gerstenkorn, S. (1972) AA, 18, 209 Neufeld, M. J. (1995) The Rocket and the Reich. The Smithsonian Institution. pp. 80–1, 156, 172. ISBN 0-674-77650-X.

18. Papadimitriou, C. H. (1995) Computational Complexity. Addison-Wesley, Reading Massachusetts. Press W. H., Teukolsky S. A., Vetterling W. T., Flannery B. P. (1995) Numerical Recipes in C, 2nd edn. Cambridge Univ. Press, New York

19. Rojas, G., Gregorio-Hetem, J., Hetem, A. (2008) MNRAS, 387, Issue 3, pp. 1335-1343. Sutton, G.P., & Biblarz, O. (2001) Rocket Propulsion Elements 7th editon, JOHN WILEY & SONS, INC.

20. Torres C. A. O. (1998) Publicação Especial do Observatório Nacional, No. 10/99. Observatório Nacional, Rio de Janeiro Torres C. A. O.,

21. Quast G. R., de la Reza R., Gregorio-Hetem J., Lépine J. R. D. (1995) AJ, 109, 2146 van den Ancker M. E., Meeus G., Cami J., Waters L. B. F. M.,Waelkens C. (2001) A&A, 369, 217

22. Wilking B. A., Lada C. J., Young E. T. (1989) ApJ, 340, 823 Young, W. C. (1989) Roark's formulas for stress and strain, McGraw-Hill

Chapter 5

FUSION OF VISUAL AND THERMAL IMAGES USING GENETIC ALGORITHMS

Sertan Erkanli[1,2], Jiang Li[2] and Ender Oguslu[1,2]

[1]Turkish Air Force Academy, Turkey
[2]Old Dominion University, USA

INTRODUCTION

Biometric technologies such as fingerprint, hand geometry, face and iris recognition are widely used to identify a person's identity. The face recognition system is currently one of the most important biometric technologies, which identifies a person by comparing individually acquired face images with a set of pre-stored face templates in a database. Though the human perception system can identify faces relatively easily, face reorganization using computer techniques is challenging and remains an active research field. Illumination and pose variations are currently the two obstacles limiting performances of face recognition systems. Various techniques have been proposed to overcome those limitations in recent years. For instance, a three dimensional face recognition system has been investigated to solve the illumination and pose variations simultaneously [Bowyer et al., 2004; S. Mdhani et al., 2006]. The illumination variation problem can also be mitigated by additional sources such as infrared (IR) images [D. A. Socolinsky & A. Selinger, 2002]. Thermal face recognition systems have received little attention in comparison with recognition in visible spectra partially due to the high cost associated with IR cameras. Recent technological advances of IR cameras make it practical for face recognition. While thermal face recognition systems are advantageous for detecting disguised faces or when there is no control over illumination, it is challenging to recognize faces in IR images because 1) it is difficult to segment faces from background in low resolution IR images and 2) intensity values in IR images are not consistent due to the fact that different body temperatures result in different intensity values in IR images. The overall goal of this research is to develop computational methods for obtaining efficiently improved images. The research objective will be accomplished by integrating enhanced visual images with IR Images through the following steps:

1)   Enhance optical images,

2)   Register the enhanced optical images with IR images, and

3)   Fuse the optical and IR images with the help of Genetic Algorithm.

Section 2 surveys related work for IR imaging, image enhancement, image registration and image fusion. Section 3 discusses the proposed nonlinear image enhancement methods.

Section 4 presents the proposed image fusion algorithm.

Section 5 reports the experimental results of the proposed algorithm. Section 6 concludes this research.

LITERATURE SURVEY

In this section, we will present related work in IR Image technology, nonlinear image enhancement algorithms, image registration and image fusion

IR TECHNOLOGY

One type of electromagnetic radiation that has received a lot of attention recently is Infrared (IR) radiation. IR refers to the region beyond the red end of the visible color spectrum, a region located between the visible and the microwave regions of the electromagnetic spectrum. Today, infrared technology has many exciting and useful applications. In the field of infrared astronomy, new and fascinating discoveries are being made about the Universe and medical imaging as a diagnostic tool. Humans, at normal body temperature, radiate most strongly in the infrared, at a wavelength of about 10 microns. The area of the skin that is directly above a blood vessel is, on average, 0.1 degrees Celsius warmer than the adjacent skin. Moreover, the temperature variation for a typical human face is in the range of about 8 degrees Celsius [F. Prokoski, 2000]. In fact, variations among images from the same face due to changes in illumination, viewing direction, facial expressions, and pose are typically larger than variations introduced when different faces are considered. Thermal IR imagery is invariant to variations introduced by illumination facial expressions since it captures the anatomical information. However, thermal imaging has limitations in identifying a person wearing glasses because glass is a material of low emissivity, or when the thermal characteristics of a face have changed due to increased body temperature (e.g., physical exercise) [G. S. Kong et al., 2005]. Combining the IR and visual techniques will benefit face detection and recognition.

NONLINEAR IMAGE ENHANCEMENT TECHNIQUES

The nonlinear log transform

The non-linear log transform converts an original image g into an adjusted image g′ by applying the log function to each pixel g[m, n] in the image,

$$g'[m, n] = k\log(g[m, n])$$

where k=L/log(L) is a scaling factor that preserve the dynamic range and L is intensity. The log transform is typically applied either to dark images where the overall contrast is low, or to images that contain specular reflections or glints. In the former case, the brightening of the dark pixels leads to an overall increase in brightness. In the latter case, the glints are suppressed thus increasing the effective dynamic range of the image. The log function as defined in equation 1 is not parameterized, i.e. it is a single input/output transfer function. A modified parameterized function was proposed by Schreiber in [W. F. Schreiber, 1978] as: image,

$$g'(l) = (L-1)\left[\frac{\log(1+\alpha g(l)) - \log(\alpha+1)}{\log(1+\alpha L) - \log(\alpha+1)}\right] + 1$$

$$(2)$$

where α parameterizes the non-linear transfer function.

REGISTRATION

Image registration is a basic task in image processing to align two or more images, usually refereed as a reference, and a sensed image [R. C. Gonzalez et al., 2004]. Registration is typically a required process in remote sensing [L. M. G. Fonseca & B. S. Manjunath, 1996], medicine and computer vision. Registration can be classified into four main categories according to the manner how the image is obtained [B. Zitova & J. Flusser, 2003]:

- Different view points: Images of the same scene taken from different viewpoints.
- Different times : Images of the same scene taken at different times.
- Different sensors : Images of the same scene taken by different sensors.
- Scene to model registration : Images of a scene taken by sensors and images of the same scene but from a model (digital elevation model).

It is impossible to implement a comprehensive method useable to all registration tasks and there are many different registration algorithms. The focus is on the feature based registration techniques in this research and they usually consist of the following three steps [B. Zitova & J. Flusser, 2003].

Feature detection: The step tries to locate a set of control points such as edges, line intersections and corners in the image. They could be manually or automatically detected.

- Feature matching: The second step is to establish the correspondence between the features detected in the sensed image and those detected in the reference image.

- Transform model estimation, Image resampling and Geometric transformation: The sensed image is transformed and resampled to match the reference image by proper interpolation techniques [B. Zitova & J. Flusser, 2003].

Each registration step has its specific problems. In the first step, features that can be used for registration must spread over the images and be easily detectable. The determined feature sets in the reference and sensed images must have enough common elements, even though the both images do not cover exactly the same scene. Ideally, the algorithm should be able to detect the same features [B. Zitova & J. Flusser, 2003]. In the second step, known as feature matching, physically corresponded features can be dissimilar because of the different imaging conditions and/or the different spectral sensitivities of the sensors. The choice of the feature description and measuring of similarity has to take into account of these factors. The feature descriptors should be efficient and invariant to the assumed degradations. The matching algorithm should be robust and efficient. Single features without corresponding counterparts in the other image should not affect its performance [B. Zitova & J. Flusser, 2003]. In the last step, the selection of an appropriate resampling technique is restricted by the trade-off between the interpolation accuracy and the computational complexity. In the literature, there are popular techniques such as the nearest-neighbor and bilinear interpolation [B. Zitova & J. Flusser, 2003]

Genetic Algorithm

Optimization can be distinguished by either discrete or continuous variables. Discrete variables have only a finite number of possible values, whereas continuous variables have an infinite number of possible ones. Discrete variable optimization is also known as combinatorial optimization, because the optimum solution consists of a certain combination of variables from the finite pool of all possible variables. However, when trying to find the minimum value of f(x) on a number line, it is more appropriate to view the problem as continuous [J. H. Holland, 1975; S. K. Mitra et al., 1998]. Genetic algorithms manipulate a population of potential solutions for the problem to be solved. Usually, each solution is coded as a binary string, equivalent to the genetic material of individuals in nature.

Each solution is associated with a fitness value that reflects how good it is, compared with other solutions in the population. The higher the fitness value of an individual, the higher its chances of survival and reproduction in the subsequent generation. Recombination of genetic material in genetic algorithms is simulated through a crossover mechanism that exchanges portions between strings. Another operation, called mutation, causes sporadic and random alteration of the bits in strings. Mutation has a direct analogy in nature and plays the role of regenerating lost genetic material [M. Srinivas & L. M. Patnaik, 1994]. GAs have found applications in many fields including image processing [J. Zhang , 2008; L. Yu et al., 2008].

Continuous Genetic Algorithm (CGA)

GAs typically represent solution as binary strings. For many applications, it is more convenient to denote solutions as real numbers known as continuous Genetic algorithms (CGA). CGAs have the advantage of requiring less storage and are faster than the binary counterparts. Figureure 1 shows the flowchart of simple CGA [Randy L. Haupt & Sue Ellen Haupt, 2004].

Components of a Continuous Genetic Algorithm

The various elements in the flowchart are described below [D.Patnaik, 2006].

Cost function

The goal of GAs is to solve an optimization problem defined as a cost function with a set of parameters involved. In CCA, the parameters are organized as a vector known as a chromosome. If the chromosome has N_{var} variables (an N-dimensional optimization problem) given by $P_1, P_2, P_3 \quad P_{var}$ then the chromosome is written as an array with 1x N_{var} elements as [Randy L. Haupt & Sue Ellen Haupt, 2004]:

$$\text{chromosome} = [p_1, p_2, p_3, \ldots, p_{N_{var}}]$$

$$(3)$$

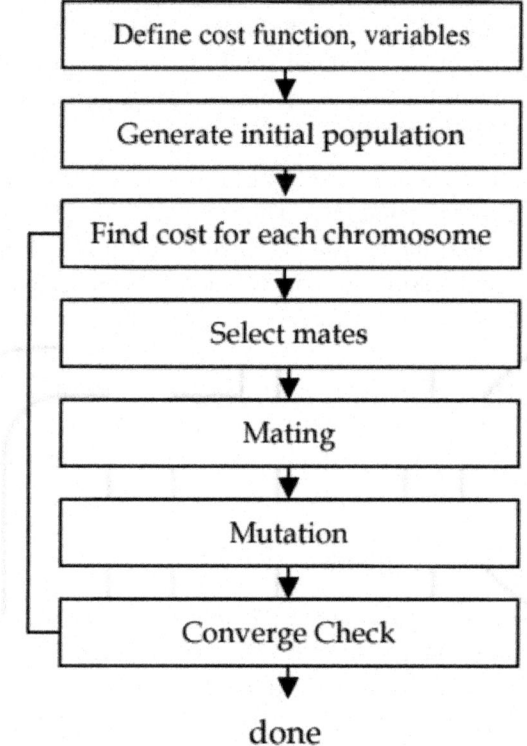

Figure.1: Flowchart of CGA

In this case, the variable values are represented as floating-point numbers. Each chromosome has a cost found by evaluating the cost function f at the variables $p_1, p_2, p_3, ..., p_{N_{var}}$,

$$\text{cost} = f(\text{chromosome}) = f(p_1, p_2, p_3, ..., p_{N_{var}})$$

(4)

Equations (3) and (4) along with applicable constraints constitute the problem to be solved. Since the GA is a search technique, it must be limited to exploring a reasonable region of variable space. Sometimes this is done by imposing a constraint on the problem. If one does not know the initial search region, there must be enough diversity in the initial population to explore a reasonably sized variable space before focusing on the most promising regions.

INITIAL POPULATION

To begin the CGA process, an initial population of N_{pop} must be defined, a matrix represents the population, with each row being a $1 \times N_{var}$ chromosome of continuous values [D.$_{\text{Patnaik}}$, 2006]. Given an initial population of N $_{\text{pop}}$ chromosomes, the full matrix of $N_{\text{pop,}}$ N_{var} random values is generated by:

$$pop = rand(N_{pop}, N_{var})$$

$$(5)$$

All variables are normalized to have values between 0 and 1. If the range of values is between lo p and $^{p_{lo}}$ $^{and\,p_{hi}}$, then the normalized values are given by

$$p = (p_{hi} - p_{lo})p_{norm} + p_{lo}$$

$$(6)$$

p_{lo} = highest number in the variable range
p_{hi} = lowest number in the variable range
p_{norm} = normalized value of variable

This society of chromosomes is not a democracy: the individual chromosomes are not all created equal. Each one's worth is assessed by the cost function. So at this point, the chromosomes are passed to the cost function for evaluation [Randy L. Haupt & Sue Ellen Haupt, 2004]. Now is the time to decide which chromosomes in the initial population are good enough to survive and possibly reproduce offspring in the next generation. As done for the binary version of the algorithm, the N_{pop} costs and associated chromosomes are ranked from lowest cost to highest cost. This process of natural selection occurs in each iteration to allow the population of chromosomes to evolve. Of the N_{pop} chromosomes in a given generation, only the top N_{keep} are kept for mating and the rest are discarded to make room for the new offspring [Randy L. Haupt & Sue Ellen Haupt, 2004].

Pairing

A set of eligible chromosomes is randomly selected as parents to generate next generation. Each pair produces two offspring that contain traits from each parent. The more similar the two parents, the more likely are the offspring to carry the traits of the parents.

Mating

As for the binary algorithm, two parents are chosen to produce offsprings. Many different approaches have been tried for crossing over in continuous GAs. The simplest method is to mark a crossover points first, then parents exchange their elements between the marked crossover points in the chromosomes. Consider two parents:

$$parent_1 = [p_{m1}, ..., p_{mN_{var}}]$$

$$parent_2 = [p_{d1}, ..., p_{dN_{var}}]$$

(7)

two offspring's might be produced as:

$$offspring_1 = [p_{m1}, p_{m2}, p_{d3}, p_{d4}, p_{m5}, p_{m6}, ..., p_{mN_{var}}]$$

$$offspring_2 = [p_{d1}, p_{d2}, p_{m3}, p_{m4}, p_{d5}, p_{d6}, ..., p_{dN_{var}}]$$

(8)

Natural selection

The extreme case is selecting N_{var} points and randomly choosing which of the two parents will contribute its variable at each position. Thus one goes down the line of the chromosomes and, at each variable, randomly chooses whether or not to swap information between the two parents. This method is called uniform crossover [Randy L. Haupt & Sue Ellen Haupt, 2004]:

$$offspring_1 = [p_{m1}, p_{d2}, p_{d3}, p_{d4}, p_{d5}, p_{m6}, ..., p_{dN_{var}}]$$

$$offspring_2 = [p_{d1}, p_{m2}, p_{m3}, p_{m4}, p_{m5}, p_{d6}, ..., p_{mN_{var}}]$$

(9)

The problem with these point crossover methods is that no new information is introduced: each continuous value that was randomly initiated in the initial population is propagated to the next generation, only in different combinations. Although this strategy worked fine for binary representations, in case of continuous variables, we are merely interchanging two data points. These approaches totally rely on mutation to introduce new genetic material. The blending methods remedy this problem by finding ways to combine variable values from the two parents into new variable values in the offspring [Randy L. Haupt & Sue Ellen Haupt, 2004]. A single offspring variable value, pnew, comes from a combination of the two corresponding offspring variable values:

$$pnew = \beta p_{mn} + (1 - \beta) p_{dn}$$

(10)

Where

β = random number in the interval [0, 1]

p_{mn} = the nth variable in the mother chromosome

p_{dn} = the nth variable in the father chromosome

The same variable of the second offspring is merely the complement of the first (i.e., replacing β by 1- p). If β =1, then p_{mn} propagates in its entirety and p_{dn} ,, dies. In contrast, if β = 0, then p_{dn} propagates in its entirety and p_{mn}

dies. When β = 0.5, the result is an average of the variables of the two parents. This method is demonstrated to work well on several interesting problems in [Randy L. Haupt & Sue Ellen Haupt, 2004].

Choosing which variables to blend is the next issue to be solved. Sometimes, this linear combination process is done for all variables to the right or to the left of some crossover point. Any number of points can be chosen to blend, up to N_{var} values where all variables are linear combinations of those of the two parents. The variables can be blended by using the same β for each variable or by choosing different β 's for each variable. These blending methods effectively combine the information from the two parents and choose values of the variables between the values bracketed by the parents; however, they do not allow introduction of values beyond the extremes already represented in the population. The simplest way is the linear crossover [Randy L. Haupt & Sue Ellen Haupt, 2004], where three offspring are generated from two parents by

$$pnew_1 = 0.5p_{mn} + 0.5p_{dn}$$
$$pnew_2 = 1.5p_{mn} - 0.5p_{dn}$$
$$pnew_3 = -0.5p_{mn} + 1.5p_{dn}$$

(11)

Any variable outside the bounds is discarded. Then the best two offspring are chosen to propagate. Of course, the factor 0.5 is not the only one that can be used in such a method. Heuristic crossover [Randy L. Haupt & Sue Ellen Haupt, 2004] is a variation where some random number, β , is chosen on the interval [0, 1] and the variables of the offspring are formed by

$$pnew = \beta(p_{mn} - p_{dn}) + p_{mn}$$

(12)

Variations on this theme include choosing any number of variables to modify and generate different β for each variable. This method also allows generations of offspring outside the value ranges of the two parent variables. If this happens, the offspring is discarded and the algorithm tries to use another b. The blend crossover (BLX-α) method [Randy L. Haupt & Sue Ellen Haupt, 2004] begins by choosing some parameters that determine the distance outside the bounds of the two parent variables that the offspring variable may lay. This method allows new values outside of the range of the parents without letting the algorithm stray too far.

The algorithm is a combination of an extrapolation method with a crossover method. The goal was to find a way to closely mimic the advantages of the binary GA mating scheme. It begins by randomly selecting a variable in the first pair of parents to be the crossover point:

$$\alpha = roundup\{random * N_{var}\}$$

(14)

where

$$parent_1 = [p_{m1}, ..., p_{m\alpha}, ..., p_{mN_{var}}]$$
$$parent_2 = [p_{d1}, ..., p_{d\alpha}, ..., p_{dN_{var}}]$$

(14)

where the m and d subscripts discriminate between the mom and the dad parent. Then the selected variables are combined to form new variables that will appear in the children:

$$pnew_1 = p_{m\alpha} - \beta[P_{m\alpha} - P_{d\alpha}]$$
$$pnew_2 = p_{d\alpha} + \beta[P_{m\alpha} - P_{d\alpha}]$$

(15)

where β is a random value between 0 and 1. The final step is to complete the crossover with the rest of chromosome:

$$offspring_1 = [p_{m1}, p_{m2}, \cdot p_{new1} ..., P_{dN_{var}}]$$
$$offspring_2 = [p_{d1}, p_{d2}, \cdot p_{new2} ..., P_{mN_{var}}]$$

(16)

where β is also a random value between 0 and 1. The final is to complete the crossover with the rest of the chromosome as before: If the first variable of the chromosomes is selected, then only the variables to the right of the selected variable are swapped. If the last variable of the chromosomes is selected, then only the variables to the left of the selected variable are swapped. This method does not allow offspring variables outside the bounds set by the parent unless $\beta > 1$.

Mutation

If care is not taken, the GA can converge too quickly into one region on the cost surface. If this area is in the region of the global minimum, there is no problem. However, some functions have many local minima. To avoid overly fast convergence, other areas on the cost surface must be explored by randomly introducing changes, or mutations, in some of the variables. Random numbers are used to select the row and columns of the variables that are to be mutated [Randy L. Haupt & Sue Ellen Haupt, 2004].

Next generation

After all these steps, the chromosomes in the starting population are ranked and the bottom ranked chromosomes are replaced by offspring from the top ranked parents to produce the next generation. Some random variables are selected for mutation from the bottom ranked chromosomes. The chromosomes are then ranked from lowest cost to highest cost. The process is iterated until a global solution is achieved.

Image fusion

In last decades, the rapid developments of image sensing technologies make multisensory systems popular in many applications. Researchers have begun to work on the fields of these systems such as medical imaging, remote sensing and the military applications [D.Patnaik, 2006]. The outcome of using these techniques is a great increase of the amount of diversity data available. Multi-sensor image data often present complementary information about the region surveyed so that image fusion provides an effective method to enable comparison and analysis of such data [H. Wang, 2004]. Image fusion is defined as the process of combining information in two or more images of a scene to enhance viewing or understanding of the scene. The fusion process must preserve all relevant information in the fused image [A. Mumtaz & A. Majid, 2008; S. Erkanli & Zia-Ur Rahman, 2010]. Image fusion can be done at pixel, feature and decision levels. Out of these, the pixel level fusion method is the simplest technique, where average/weighted averages of individual pixel intensities are taken to construct a fused image [K. Kannan & S. Perumal, 2007]. Despite their simplicity, these methods are not used nowadays because of some serious disadvantages they possess. For instance, the contrast of the fused information is reduced and also redundant information is introduced in the fused image, which may mask the useful information. These disadvantages are overcomed by feature level and decision level fusion methods. Feature and decision level fusion methods are based on human vision system. Decision level fusion combines the results from multiple algorithms to yield a final fused image. Several pyramid transform methods for feature level fusion have been suggested [A. Wang et al., 2006]. Recently, developed methods based on the wavelet transform become popular [A. Wang et al., 2006]. In the method source images are decomposed into subimages of different resolutions and in each subimage different features become prominent. To fuse the original source images, the corresponding subimages of different source images are combined based some criteria to form composite subimages. Inverse pyramid transform of composite transform gives the final fused image

ENHANCING POOR VISIBILITY IMAGES

The human visual system (HVS) allows individuals to assimilate information from their environment [S. Erkanli & Zia-Ur Rahman, 2010b; H. Kolb, 2003]. The HVS perceives colors and detail across a wide range of photometric intensity levels much better than electronic cameras. The perceived color of an object, additionally, is almost independent of the type of illumination, i.e., the HVS is color constant. Electronic cameras suffer, by comparison, from limited dynamic range and the lack of color constancy and current imaging and display devices such as CRT monitors and printers have limited dynamic range of about two orders of magnitude, while the best photographic prints can provide contrast up to $10^3 : 1$. However; real world scenes can have a dynamic range of six orders of magnitude [S. Erkanli & Zia-Ur Rahman, 2010b; L. Tao et al., 2005]. This can result in overexposure that causes saturation in high contrast images, or underexposure in dark images [Z. Rahman, 1996]. The idea behind enhancement techniques are to bring out details in images that are otherwise too dim to be perceived either due to insufficient brightness or insufficient contrast [Z. Rahman, 1997]. A large number of image enhancement methods have been developed, like log transformations, power law transformations, piecewise-linear transformations and histogram equalization. However these enhancement techniques are based on global processing which results in a single mapping between the input and the output intensity space. These techniques are thus not sufficiently powerful to handle images that have both very bright and very dark regions. Other image enhancement techniques are local in nature, i.e., the output value depends not only on the input pixel value but also on pixel values in the neighborhood of the pixel. These techniques are able to improve local contrast under various illumination conditions.

Single-Scale Retinex (SSR), is a modification of the Retinex algorithm introduced by Edwin Land [G. D. Hines et al., 2004; E. Land, 1986]. It provides dynamic range compression (DRC), color constancy, and tonal rendition. SSR gives good results for DRC or tonal rendition but does not provide both simultaneously. Therefore, the Multi-Scale Retinex (MSR) was developed by Rahman et al. The MSR combines several SSR outputs with different scale constants to produce a single output image, which has good DRC, color constancy and good tonal rendition. The outputs of MSR display most of the detail in the dark pixels but at the cost of enhancing the noise in these pixels and the tonal rendition is poor in large regions of slowly changing intensity. As a result, Multi-Scale Retinex with Color Restoration (MSRCR) was developed by Jobson et al., for synthesizing local contrast improvement, color constancy and lightness/color rendition. Other non-linear enhancement models include the Illuminance Reflectance Model for Enhancement (IRME) proposed by Tao et al. [L. Tao et al., 2005], and the Adaptive and Integrated Neighborhood-Dependent Approach for Nonlinear Enhancement (AINDANE) described by Tao [L.Tao, 2005]. Both use a nonlinear function for luminance enhancement and tune the intensity of each pixel based on its relative magnitude with respect to the neighboring pixels. In this section, a new image enhancement approach is described: Enhancement Technique for Nonuniform and Uniform-Dark Images (ETNUD). The details of the new algorithm are given in Section 3.2, respectively. Sections 3.3 describe experimental results and compare our results with other techniques for image enhancement. Finally in Section 3.4, conclusions are presented.

Enhancement Technique for Nonuniform and Uniform-Dark Images (ETNUD)

The major innovation in ETNUD is in the selection of the transformation parameters for DRC, and the surround scale and color restoration parameters. The following sections describe the selection mechanisms.

Selection of transformation parameters for DRC

The intensity I of the color image Ic can be determined by:

$$I(m,n) = 0.2989r(m,n) + 0.587g(m,n) + 0.114b(m,n)$$

(17)

where r, g, b are the red, green, and blue components of I_c respectively, and m and n are the row and column pixel locations respectively. Assuming I to be 8-bits per pixel, I_n is the normalized version of I, such that:

$$I_n(m,n) = I(m,n) / 255$$

(18)

Using linear input-output intensity relationships typically does not produce a good visual representation compared with direct viewing of the scene. Therefore, nonlinear transformation for DRC is used, which is based on some information extracted from the image histogram. To do this, the histogram of the intensity images is subdivided into four ranges:

$r_1 = 0$–63, $r_2 = 64$–127, $r_3 = 128$–191 and $r_4 = 192$–255. I_n is mapped to I_n^{drc} using the following:

$$I_n^{drc} = \begin{cases} (I_n)^x + \alpha & 0 < x < 1 \\ (0.5 + (0.5I_n)^x) + \alpha & x \geq 1 \end{cases}$$

(19)

The first mapping pulls out the details in the dark regions, and the second suppresses the bright overshoots. The value of x is given by

$$x = \begin{cases} 0.2, & if \ (f(r_1 + r_2) \geq f(r_3 + r_4)) \wedge (f(r_1) \geq f(r_2)) \\ 0.5, & if \ (f(r_1 + r_2) \geq f(r_3 + r_4)) \wedge (f(r_1) < f(r_2)) \\ 3.0, & if \ (f(r_1 + r_2) < f(r_3 + r_4)) \wedge (f(r_3) \geq f(r_4)) \\ 5.0, & if \ (f(r_1 + r_2) < f(r_3 + r_4)) \wedge (f(r_3) < f(r_4)) \end{cases}$$

(20)

where f ( ) a refers to number of pixels between the range (a), $f(a_1 + a_2) = f(a_1) + f(a_2)$, and $\wedge$ is the logical AND operator. α is the offset parameter, helping to adjust the brightness of image. The determination of the x values and their association with the range-relationships as given in Equation 20 was done experimentally using a large number of non-uniform and uniform dark images and x value can be also determined manually. The DRC mapping of the intensity image performs a visually dramatic transformation. However, it tends to have poor contrast, so a local, pixel dependent contrast enhancement method is used to improve the contrast.

Selection of surround parameter and color restoration

Many local enhancement methods rely on center/surround ratios [L. Tao, 2005]. Hurlbert [A. C. Hulbert, 1989] investigated the Gaussian as the optimal surround function. Other surround functions proposed by [E. Land, 1986] were compared with the performance of the Gaussian proposed by [D. J. Jobson, et al., 1997]. Both investigations determined that the Gaussian form produced good dynamic range compression over a range of space constants. Therefore, the luminance information of surrounding pixels is obtained by using 2D discrete spatial convolution with a Gaussian kernel, G(m, n) defined as:

$$G(m,n) = K \exp\left(-\frac{m^2 + n^2}{\sigma_s^2}\right)$$

(21)

where σ s is the surround space constant equal to the standard deviation of G(m, n), and K is determined under the constraint that $\sum_{m,n} G(m,n) = 1$. The center-surround contrast enhancement is defined as:

$$I_{enh}(m,n) = 255(I_n^{drc}(m,n))^{E(m,n)}$$

(22)

where, E(m, n) is given by:

$$E(m,n) = \left[\frac{I_{filt}(m,n)}{I(m,n)} \right]^{S}$$

(23)

Where

$$I_{filt}(m,n) = I(m,n) * G(m,n)$$

(24)

S is an adaptive contrast enhancement parameter related to the global standard deviation of the input intensity image, I(m, n), and '*' is the convolution operator, I(m, n) is defined by:

$$S = \begin{cases} 3 & for & \sigma \le 7 \\ 1.5 & for & 7 < \sigma \le 20 \\ 1 & for & \sigma \ge 20 \end{cases}$$

(25)

σ is the contrast—standard deviation—of the original intensity image. If σ < 7, the image has poor contrast and the contrast of the image will be increased. If σ ≥ 20, the image has sufficient contrast and the contrast will not be changed. Finally, the enhanced image can be obtained by linear color restoration based on chromatic information contained in the original image as:

$$S_j(x,y) = I_{enh}(x,y) \frac{I_j(x,y)}{I(x,y)} \lambda_j$$

(26)

where $j \in \{r,g,b\}$ represents the RGB spectral band and j_λ is a parameter which adjusts the color hue.

Evaluation citeria

In this work, following evaluation criteria was used.

A new metric

There are some metrics such as brightness and contrast to characterize an image. Another such metric is sharpness. Sharpness is directly proportional to the high-frequency content of an image. So the new metric is defined as [Z. Rahman, 2009]:

$$S = \sqrt{\|h \otimes I\|^2} = \sqrt{\sum_{v_1=0}^{M_1-1} \sum_{v_2=0}^{M_2-1} \left| \hat{h}[v_1, v_2] \hat{I}[v_1, v_2] \right|}$$

(27)

where h is a high-pass filter, periodic with period $M_1 x M_2$, and $\hat{h}$ is its direct Discrete Fourier Transform (DFT). I is also DFT of Image I. The role of $\hat{h}$ (or h) is to weight the energy at the high frequencies relative to the low frequencies, thereby emphasizing the contribution of the high frequencies to S. The larger the value of S, the greater the sharpness of I and conversely. Equation 27 defines how the sharpness should be computed and defined as:

$$\hat{h}[v_1, v_2] = 1 - \exp\left(-\frac{v_1^2 + v_2^2}{\sigma^2} \right)$$

(28)

where σ is the parameter at which the attenuation coefficient $= 1.0 - e^{-1} \approx 2/3$. A smaller value of σ implies that fewer frequencies are attenuated and vice versa. For this research $\sigma = 0.15$.

Image qality asessment

The overall quality of images can be measured by using the brightness μ, contrast σ and sharpness S, where brightness and contrast are assumed to be the mean and the standard deviation. However, instead of using global statistics, it is used regional statistics. In order to do this [Z. Rahman, 2009]:

1. Divide the $M_1 x M_2$ image I into $(M_1/10)x(M_2/10)$ non-overlapping blocks, I_i, $i=1,...,100$, such that $I \approx \cup_{i=1}^{N} I_i$, (Total Number of Regions are 100).
2. For each block compute the measures, μ, σ and S,
3. Classify the block as either GOOD or POOR based on the computed measure (will be discussed with the following).
4. Classify the image as a whole as GOOD or POOR based upon the classification of regions (will be discussed with the following). The following criteria are used for brightness, contrast and sharpness [Z. Rahman, 2009]

   1. Let μ_n be normalized brightness parameter, such that:

$$\mu_n = \begin{cases} \mu / 255 & \mu < 154 \\ 1 - \mu / 255 & otherwise \end{cases}$$

(29)

A region is considered to have sufficient brightness when $0.4 \le \mu_n \le 0.6$.

Let σ_n be normalized contrast parameter, such that:

$$\sigma_n = \begin{cases} \sigma/128 & \mu \le 64 \\ 1 - \sigma/128 & otherwise \end{cases}$$

(30)

A region is considered to have sufficient contrast when $0.25 \le \sigma_n \le 0.5$. When $\sigma_n < 0.25$, the region has poor contrast, and when $\sigma_n > 0.5$, the region has too much contrast.

Let S_n be normalized sharpness parameter, such that $S_n = min(2.0, S/100)$. When $S_n > 0.8$, the region has sufficient sharpness. Image Quality is evaluated using by

$$Q = 0.5\mu_n + \sigma_n + 0.1S_n$$

(31)

where $0 < Q < 1.0$ is the quality factor. A region is classified as good when $Q > 0.55$, , and poor when $\sigma_n \le 0.5$. image is classified as GOOD when the total number of regions classified as GOOD, GOOD, $N_G > 0.6N$. .

Experimental result

The image samples for ETNUD were selected to be as diverse as possible so that the result would be as general as possible. MATLAB was used for AINDANE and IRME algorithms and their codes were developed by the author and research team. MSRCR enhancement was done with commercial software, Photo Flair. From visual experience, the following statements are made about the proposed algorithm:

1. In the Luminance enhancement part it has been shown that ETNUD works well for darker images and the technique adjusts itself to the image (Figureure 2).

2. In the contrast enhancement part it is clear that unseen or barely seen features of low contrast images are made visible.

3. In Figureure 2 Gamma Correction with $\gamma = 1.4$ does not provide good visual enhancement. IRME and MSRCR bring out the details in the dark but have some enhancement of noise in the dark regions, which can be considered objectionable. AINDANE does not bring out the finer details of the image. The ETNUD algorithm gives good result and outperforms the other algorithms if the results are compared (in Table 1) due to the Evaluation Criteria. The ETNUD provides better visibility enhancement the best sharpness can be adjusted by the α parameter in Equation 19.

Table 1: The Results of Evaluation Criteria for Figure 2.

Figure 2	Original Image	Gamma	Irme	Aindane	Msr	Etnud
Number of Good Regions	32	52	95	90	90	99
Number of Poor Regions	68	48	5	10	10	1

CONCLUSION

The ETNUD image enhancement algorithms provide high color accuracy and better balance between the luminance and contrast in images.

ENTROPY-BASED IMAGE FUSION WITH CONTINUOUS GENETIC ALGORITHM

Image fusion is defined as the process of combining information from two or more images of a scene to enhance the viewing or understanding of that scene. The images that are to be fused can come from different sensors, or have been acquired at different times, or from different locations. Hence, the first step in any image fusion process is the accurate registration of the image data. This is relatively straightforward if parameters such as the instantaneous field-of-view (IFOV), and locations and orientations from which the images are acquired are known, especially when the sensor modalities produce images that use the same coordinate space. This is more of a challenge when sensor modalities differ significantly and registration can only be accomplished at the information level. Hence, the goal of the fusion process is to preserve all relevant information in the component images and place it in the fused image (FI). This requires that the process minimize the noise and other artifacts in the FI. Because of this, the fusion process can be also regarded as an optimization problem [K. Kannan and S. Perumal, 2002]. In recent years, image fusion has been applied to a number of diverse areas such as remote sensing [T. A. Wilson, and S. K. Rogers,1997], medical imaging [C. S. Pattichis and M. S. Pattichis, 2001], and military applications [B. V. Dasarathy, 2002].

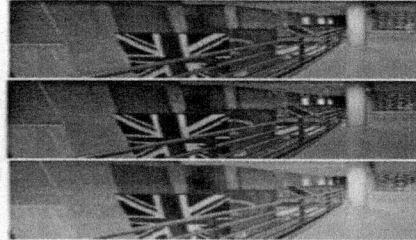

Figure. 2: Comparisons of Enhancement Techniques: (top-left) Original; (top-right) IRME; (middle-left) Gamma correction, g = 1.4; (middle-right) MSR; (bottom-left) AINDANE;(bottom-right) ETNUD.

Image fusion can be divided into three processing levels: pixel, feature and decision. These methods increase in abstraction from pixel to feature to decision levels. In the pixel-level approach, simple arithmetic rules like average of individual pixel intensities or more sophisticated combination schemes are used to construct the fused image. At the featurelevel, the image is classified into regions with known labels, and these labeled regions from different sensor modalities are used to combine the data. At the decision level, a combination of rules can be used to include part of the data or not.

Genetic algorithms (GA) are an optimization technique that seeks the optimum solution of a function based on the Darwinian principles of biological evolution. Even though there are several methods of performing and evaluating image fusion, there are still many open questions. In this section, a new measure of image fusion quality is provided and compared with many existing ones. The focus is on pixel-level image fusion (PLIF) and a new image fusion technique that uses GA is proposed. The GA is used to optimize the parameters of the fusion process to produce an FI that contains more information than either of the individual images. The main purpose of this section is in finding the optimum weights that are used to fuse images with the help of CGA. The techniques for GA and image fusion are given in Section 4.2. Section 4.3 describes the evaluation criteria. Section 4.4 describes the experimental results, and compares our results with other image fusion techniques. In Section 4.5, conclusion is provided.

The techniques of GA and image fusion

Genetic Algorithm

As stated earlier, GA is a non-linear optimization technique that seeks the optimum solution of a function via a non-exhaustive search among randomly generated solutions. GAs use multiple search points instead of searching one point at a time and attempt to find global, near-optimal solutions without getting stuck at local optima. Because of these significant advantages, GAs reduce the search time and space. However, there are disadvantages of using GAs as well: they are not generally suitable for real-time applications since the time to converge to an optimal solution cannot be predicted. The convergence time depends on the population size, and the GA crossover and mutation operators. In this fusion process, a continuous genetic algorithm has been selected.

Continuous Genetic Algorithm (CGA)

GAs typically operates on binary data. For many applications, it is more convenient to work in the analog, or continuous, data space rather than in the binary space of most GAs. Hence, CGA is used because they have the advantage of requiring less storage and are faster than binary. CGA inputs are represented by floating-point numbers over whatever range is deemed appropriate. Figureure 6 shows the flowchart of a simple CGA [Randy L. Haupt & Sue Ellen Haupt, 2004]. The various elements in the flowchart are described below:

- Definition of the cost function and the variables: The variable values are represented as floating point numbers(p_1) In each chromosome, the basic GA processing vector, there are number of value depending on the parameters $(p_1,...,p_{N\,var})$. Each chromosome has a cost determined by evaluating the cost function [Randy L. Haupt & Sue Ellen Haupt, 2004].

- ii. Initial Population: To begin the CGA process, an initial population must be defined. A matrix represents the population, with each row being a $1 \times N_{var}$ chromosome of continuous values. The chromosomes are passed to the cost function for evaluation [Randy L. Haupt & Sue Ellen Haupt, 2004].

- Natural Selection: The chromosomes are ranked from the lowest to highest cost. Of the total of chromosomes in a given generation, only the top N Keep are kept for mating and the rest are discarded to make room for the new off spring .

- Mating: Many different approaches have been tried for crossover in continuous GAs. In crossover, all the genes to the right of the crossover point are swapped. Variables are randomly selected in the first pair of parents to be the crossover point: $\alpha = (U(0,1)N_{var})$, where $U(0,1)$ is the uniform distribution. The parents are given by [Randy L. Haupt & Sue Ellen Haupt, 2004]:

$$parent_1 = [P_{m1}, ..., P_{mN_{var}}]$$

$$parent_2 = [P_{d1}, ..., P_{dN_{var}}]$$

$$(27)$$

where subscripts m and d represent the mom and dad parent. Then the selected variables are combined to form new variables that will appear in the children.

$$pnew_1 = p_{m\alpha} - \beta[P_{m\alpha} - P_{d\alpha}]$$

$$pnew_2 = p_{d\alpha} + \beta[P_{m\alpha} - P_{d\alpha}]$$

$$(33)$$

where β is a random value between 0 and 1. The final step is to complete the crossover with the rest of chromosome:

$$offspring_1 = \left[P_{m1}, P_{m2}, ..., pnew_1, ..., P_{dN_{var}} \right]$$

$$offspring_2 = \left[P_{d1}, P_{d2}, ..., pnew_2, ..., P_{mN_{var}} \right]$$

$$(34)$$

Mutation: If care is not taken, the GA can converge too quickly into one region of the cost surface. If this area is in the region of the global minimum, there is no problem. However, some functions have many local minima. To avoid overly fast convergence, other areas of the cost surface must be explored by randomly introducing changes, or mutations, in some of the variables. Multiplying the mutation rate by the total number of variables that can be mutated in the population gives the amount of mutation. Random numbers are used to select the row and columns of the variables that are to be mutated. vi. Next Generation: After all these steps, the starting population for the next generation is ranked. The bottom ranked chromosomes are discarded and replaced by offspring from the top ranked parents. Some random variables are selected for mutation from the bottom ranked chromosomes. The chromosomes are then ranked from lowest cost to highest cost. The process is iterated until a global solution is achieved [Randy L. Haupt & Sue Ellen Haupt, 2004].

Image fusion

A set of input images of a scene, captured at a different time or captured by different kinds of sensors at the same time, reveals different information about the scene. The process of extracting and combining data from a set of input images to form a new composite image with extended information content is called image fusion.

Evaluation criteria

In this section, the following criteria were defined to evaluate the performance of the image fusion algorithm.

Entropy

Entropy is often defined as the amount of information contained in an image. Mathematically, entropy is usually given as:

$$E = -\sum_{i=0}^{L-1} p_i \log_2 p_i$$

(35)

where L is the total number of grey levels, and $p = \{p_0, \ldots, p_{L-1}\}$ is the probability of occurrence of each level. An increase in entropy after fusion can be interpreted as an overall increase in the information content. Hence, one can assess the quality of fusion by assessing entropy of the original data, and the entropy of the fused data.

Mutual information indices

Mutual Information Indices are used to evaluate the correlative performances of the fused image and the source images. Let A and B be random variables with marginal probability distributions $p_A(a)$ and $p_B(b)$ and the joint probability distribution $p_{AB}(a,b)$. The mutual information is then defined as:

$$I_{AB} = \sum p_{AB}(a,b) \log[p_{AB}(a,b) / (p_A(a)p_B(b))]$$

(36)

A higher value of Mutual Information (MI) indicates that the fused image, F, contains fairly good quantity of information present in both the source images, A and B. The MI can be defined as $MI = I_{AF} + I_{BF}$.

A high value of MI does not imply that the information from the both images is symmetrically fused. Therefore, information symmetry (IS) is introduced. IS is the indication of how symmetrically distributed is the information in the fused image, with respect to input images. The higher the value of IS, the better the fusion result. IS is given by :

$$IS = 2 - abs[I_{AF} / (I_{AF} + I_{BF}) - 0.5]$$

(37)

EXPERIMENTAL RESULTS

The goal of this experiment is to fuse visual and IR images. To minimize registration issues, it is important that the visual and the thermal images are captured at the same time. Pinnacle software was used to capture the visual and the thermal images simultaneously. Although radiometric calibration is important, the thermal camera can not always be calibrated in field conditions because of constraints on time. Figure 3 shows an example where the IR and visual image were captured at the same time. It is obvious from the Figure that the images need to be registered before they can be fused since the field-of-view and the pixel resolution are obviously different. The performance of the proposed algorithm was tested and compared with different PLIF methods. The IR and visual images were not previously registered as shown in Figure 3. The registered image, base image (IR Image) and fused image with CGA are shown in Figure 4. The cost function is very simple and defined as:

$$Entropy(F = w_a V + w_b IR)$$

$$(38)$$

where V and IR are the visual and IR images, w_a and w_b are the respective associated weights, and F is the fused image. The initial population size is 100×3. The first and second columns in population matrix represent $w_a V$, and $w_b IR$ and the last column represents the cost function which is the entropy of F. Then initial population has been ranked based on the cost. In each iteration of the GA, 20 of the 100 rows are kept for mating and the rest are discarded. The crossover has been applied based on the Equation 35. The mutation rate was set to 0.20, hence the total number of mutated variables is 40. The value of a mutated variable is replaced by a new random value in the same range.

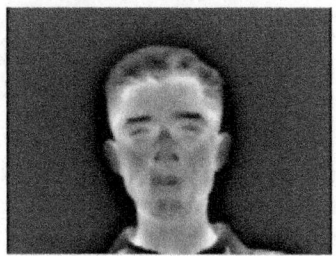

Figure. 3: Visual and IR Images: Left: Visual Image, Right: IR Image

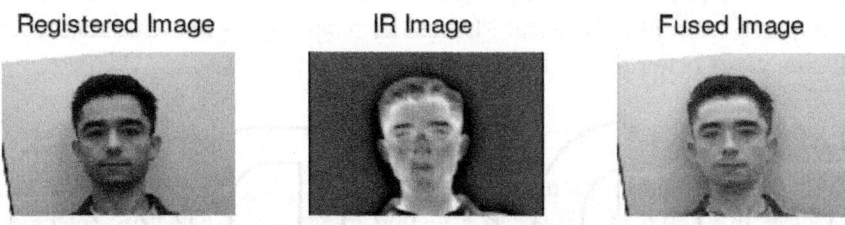

Figure. 4: The Result of Fusion: Left: Registered Images, Middle: IR Image Right: Fused Image with GA.

The CGA results after 50 iterations of the GA such that the CGA maximize the cost and find optimum weights of images. In the 2nd, 8th, and 25th iterations, the cost increased but was not associated with the global solution. The optimum solution was determined in 45th iteration and remained unchanged because it is optimum solution. Figureure 4 shows the fusion results of point-rules based PLIF. After registering IR and visual data, we determined that $w_a = 0.9931$ and $w_b = 0.0940$ provide the optimum values for maximizing the entropy cost function for the F specified in Equation 38. The evaluation of these weights results is shown in Table 2. Table 2 shows that CGA based fusion method gives better results (optimum weights for maximizing the entropy of F) for entropy and IS from which it can concluded that CGA performs better than other PLIFs.

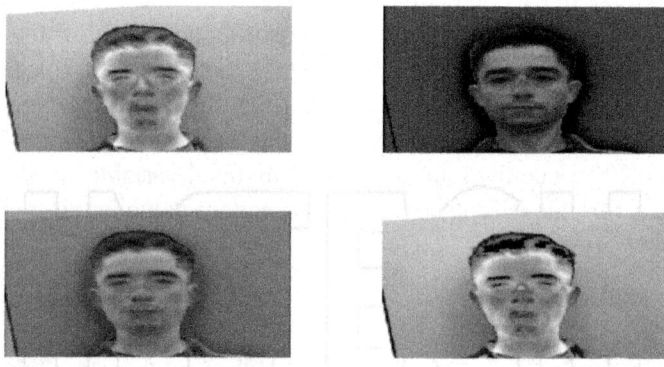

Figure. 5: Fusion Results: (top-left) highest value from IR or Visual Images; (top-right) lowest value form IR or Visual Images; (bottom-left) average of IR and Visual Images; (bottom right) threshold value.

Conclusion

In this section, CGA based image fusion algorithm was introduced and compared with other classical PLIFs. The results show that CGA based image fusion gives better result than other PLIFs.

EXPERIMENTAL RESULTS

With face recognition, a database usually exists that stores a group of human faces with known identities. In a testing image, once a face is detected, the face is cropped from the image or video as a probe to check with the database for possible matches. The matching algorithm produces a similarity measure for each of the comparing pairs. Variations among images from the same face due to changes in illumination are typically larger than variations rose from a change of face identity. In an effort to address the illumination and camera variations, a database was created, considering these variations to evaluate the proposed techniques. Besides the regular room lights, four additional spot lights are located in the front of the person that can be turned off and on in sequence to obtain face images under different illumination conditions. Note that it is important to capture visual and thermal images at the same time in order to see the variations in the facial images. Visual and thermal images are captured almost at the same time. Although radiometric calibration is important, the thermal camera can not be calibrated because of current IR camera characteristics.The Pinnacle (Pinnacle Systems Ltd.) software has been implemented to capture 16 visual and thermal images at the same time. Figureure 6 (a) and (e) shows an example of visual and thermal images taken at the same time.

Table 2: Performance Comparision of Image Fusion Methods for Figureure 4 and Figureure 5

	Highest (Fig 5)	Lowest (Fig 5)	Average (Fig 5)	Threshold (Fig 5)	GA_based (Fig 4)
Entropy	6.91	3.14	6.56	6.93	7.28
Image Quality	100	70	100	100	100
IS	1.90	1.63	1.96	1.91	1.96

In this chapter, the focus is on visual image enhancement. Then the visual images will be registered with the IR images based landmark registration algorithm. Finally, the registered IR and visual images are fused for face recognition.

Enhancement of visual images

The ETNUD algorithm was applied to 16 visual images as shown in Figureure 6 under different illumination conditions. In all Figureures besides the regular room lights, the four extra spot lights located in the front of the person were turned off and on for creating different illumination conditions. To enhance those visual images, the luminance is first balanced, then image contrast is enhanced and finally, the enhanced image is obtained by a linear color restoration based on chromatic information contained in the original image. The results in the luminance enhancement part showed that the algorithms work well for dark images. All the details, which cannot be seen in the original image, become evident. The experiment results have shown that for all color images, the proposed algorithms work sufficiently well.

IR and visual images registration

First, the IR and visual images taken from different sensors, viewpoints, times and resolution were resized for the same size. The correspondence between the features detected in the IR image and those detected in the visual image were then established. Control points were picked manually from those corners detected by the Harris corner detection algorithm from both images, where the corners were in the same positions in the two images. In the second step, a spatial transformation was computed to map the selected corners in one image to those in another image. Once the transformation was established, the image to be registered was resampled and interpolated to match the reference image. For RGB and intensity images, the bilinear or bi cubic interpolation method is recommended since they lead to better results. In the experiments, the bi cubic interpolation method was used.

Discussion

Experimental results have been applied on the database, which is created by the research team. This algorithm is categorized into four steps, which are described respectively. In the

first step, there is enhancement of visual images, as described in Section 3. The fused image should be more suitable for human visual perception and computer-processing tasks. The experience of image processing has prompted the research to consider fundamental aspects for good visual presentation of images, requiring nonlinear image enhancement techniques of visual recorded images to get a better image, which has more information from the original images. In the second step, the corners of visual and IR images were determined with the help of Harris Detection algorithm for registration purpose to use as control points. In the third step, because the source images are obtained from different sensors, they present different resolution, size and spectral characteristic, the source images have to be correctly registered. In the last step, an image fusion process is performed, which was described in Section 4. The registered images were overlapped at an appropriate transparency. The pixel value in the fused image was a weighted submission of the corresponding pixels in the IR and visual images. In the next section, results from advanced image fusion approaches are presented.

Fusion of visual and IR images

The Image fusion algorithm was applied with the help of Genetic Algorithm to the database. One of the issues is the determination of the quality of image fusion results. As part of the general theme of fusion evaluation there is a growing interest to develop methods that address the scored performance of image fusion algorithms as described in Section 4. Given the diversity of applications and various methods of evaluation metrics, there are still open questions concerning when to perform image fusion. There is an interest in exploring mean, standard deviation, entropy, mutual information, peak signal to noise ratio and image quality as described in Section 4. Because source images have different spectrum, they show quite distinct characters and have complementary information. It can be seen in Figure 6 (a and c) that the visual image does not have enough information to see the faces and is very dark. Figure 6 (b) shows that the luminance enhancement part works well for dark images and the technique adjusts itself to the image. In the contrast enhancement part it is clear that unseen or barely seen features of low contrast images were made visible. Enhancement algorithms were developed to improve the images before the fusion process. After enhancement it was found that the corners of the enhanced image and the IR image then registered the enhanced image as shown in Figure 6 (d). Then, the enhanced image was fused with the IR image in Figure 6 (f). Figure 6 show the result of CGA after 100 iterations. The optimum solution was determined with a population size of 100x3 after 76 iterations. It was determined that $w_a = 0.99$ and $w_b = 0.47$ are the optimum values for maximizing the entropy cost function which is 7.58 for the F specified in Equation 38. The evaluation of these weights results is shown in Table 3. By inspection, the faces and the details in the fused image are clearer as compared to either the original IR image or the visual image. Table 3 shows the detailed comparison results of the fused images. A is the fused image by averaging the visual and IR images. B is the fused image by the proposed approach. The total images used in this experiment were from the created database. The results show that this approach is better than the averaging fusion result.

Table 3: The Statistics of Database

Database Images	MEAN		ENTROPY		PSNR		IQ	
	A	B	A	B	A	B	A	B
1(Fig.19)	101.61	153.50	7.03	7.58	14.16	35.73	85	94
2	111.78	144.92	7.26	7.68	13.64	35.73	90	95
3	105.35	124.06	7.25	7.42	13.84	28.13	87	96
4	118.91	140.72	7.33	7.53	13.21	28.33	97	96
5	104.2	117.17	7.41	7.82	14.12	29.40	91	94
6	106.82	117.41	7.46	7.78	14.10	29.15	97	94
7	115.76	137.67	7.37	7.68	14.12	29.26	98	98
8	116.18	137.02	7.56	7.83	14.50	29.64	97	96
9	93.22	134.03	7.29	7.63	15.22	33.41	87	83
10	114.05	143.26	7.23	7.60	14.64	36.17	99	98
11	111.50	131.12	7.34	7.51	13.92	28.25	93	99
12	117.51	142.50	7.37	7.66	13.60	30.10	96	95
13	114.65	139.16	7.34	7.51	14.18	30.05	94	96
14	116.47	141.82	7.29	7.54	15.08	30.94	99	99
15	115.81	132.06	7.53	7.60	14.39	28.75	98	97
16	118.57	137.00	7.34	7.68	14.93	28.90	99	99

Figure. 6: Fusion Results for Image 1: (top-left-(a)) Original; (top-right-(b)) Enhanced; (middle left-(c)) Original; (middle-right-(d)) Enhanced; (bottom-left-(e)) IR;(bottom-right-(f)) Fused Images; Graph-Genetic Algorithm result after 100 iterations

CONCLUSIONS

In this chapter, a database for visual and thermal images was created and several techniques were developed to improve image quality as an effort to address the illumination challenge in face recognition. Firstly, one image enhancement algorithm was designed to improve the images' visual quality. Experimental results showed that the enhancement algorithm performed well and provided good results in terms of both luminance and contrast enhancement. In the luminance enhancement part, it has been shown that the proposed algorithm worked well for both dark and bright images. In the contrast enhancement part, it was proven that the proposed nonlinear transfer functions could make unseen or barely seen features in low contrast images clearly visible. Secondly, the IR and enhanced visual images taken from different sensors, viewpoints, times and resolution were registered. A correspondence between an IR and a visual image was established based on a set of image features detected by the Harris Corner detection algorithm in both images. A spatial transformation matrix was determined based on some manually chosen corners and the transformation matrix was utilized for the registration. Finally, a continuous genetic algorithm was developed for image fusion. The continuous GA has the advantage of less storage requirements than the binary GA and is inherently faster than the binary GA because the chromosomes do not have to be decoded prior to the evaluation of the cost function. Data fusion provides an integrated image from a pair of registered and enhanced visual and thermal IR images. The fused image is invariant to illumination directions and is robust under low lighting conditions. They have potentials to significantly boost the performances of face recognition systems. One of the major obstacles in face recognition using visual images is the illumination variation. This challenge can be mitigated by using infrared (IR) images. On the other hand, using IR images alone for face recognition is usually not feasible because they do not carry enough detailed information. As a remedy, a hybrid system is presented that may benefit from both visual and IR images and improve face recognition under various lighting conditions.

REFERENCES

1.   Bowyer W., Chang K. and Flynn P., A Survey of Approaches To Three-Dimensional Face Recognition, ICPR, Vol. 1, pp. 358 – 361, 2004.

2.   Dasarathy B. V., Image Fusion in the Context Of Aerospace Applications, Inform. Fusion, Vol. 3, 2002.

3.   Erkanli S. and Rahman Zia-Ur., Enhancement Technique for Uniformly and Non-Uniformly Illuminated Dark Images, ISDA 2010, Cairo, Egypt, 2010b.

4.   Erkanli S. and Rahman Zia-Ur., Wavelet Based Enhancement for Uniformly and NonUniformly Illuminated Dark Images, ISDA 2010, Cairo, Egypt, 2010c.

5.   Erkanli S.and Rahman Zia-Ur, Entropy Based Image Fusion With the help of Continuous Genetic Algorithm, IEEE ISDA Conference, December 2010.

6.   Fonseca L. M. G. and Manjunath B. S., Registration Techniques for Multisensor Remotely Sensed Imagery, Photogrammetric Engineering & Remote Sensing, Vol. 62, pp. 1049- 1056, 1996.

7.   Gonzalez R. C., Woods R. E. and Eddins S. L., Digital Image Processing, Pearson Education, Inc. Prentice Hall, 2004.

8.   Haupt Randy L. and Haupt Sue Ellen, Practical Genetic Algorithms, Second Edition, ISBN 0- 471-45565-2 Copyright © 2004

9.   John Wiley & Sons, Inc. Hines G. D., Rahman Z., Jobson D. J., and Wodell G. A., Single-Scale Retinex Using Digital Signal Processors In Proceedings Of The GSPX, 2004.

10.   Holland J. H., Adaptation In Natural and Artificial Systems, University of Michigan Press, 1975. Hulbert A. C., The Computation of Color, Ph.D. Dissertaion, Mass. Inst. Tech., Cambridge, MA, Sept. 1989.

11.   Jobson D. J., Rahman Z. and Woodell G. A., Properties and Performance of a Center/Surround Retinex, IEEE Trans.Image Processing, Vol.6, pp. 451-462, 1997.

12.   Kannan K. and Perumal S., Optimal Decomposition Level of Discrete Wavelet Transform for Pixel Based Fusion of Multi-Focused Images, International Conference On Computational Intelligence And Multimedia Applications, 2007. Kolb H., How the Retina works, American Scientist, Vol. 91, 2003.

13.   Kong G. S., Heo J., Abidi B. R., Paik J. and Abidi M. A., Recent Advances in Visual and Infrared Face recognition—a Review, Computer Vision and Image Understanding, Vol. 1, pp. 103-135, 2005.

14.   Land E., An Alternative Technique For The Computation of The Designator in The Retinex Theory of Color Vision, Proc. Of The National Academy Of Science USA, Vol. 83, pp. 2078-3080, 1986.

15.   Mdhani S., Ho J., Vetter T. and Kriegman D. J., Face Recognition Using 3-D Models: Pose and Illumination, Proceedings of the IEEE, Vol. 94, pp. 1977 –1999, 2006.

16.   Mitra S. K., Murthy C. A. and Kundu M. K., Technique for Fractal Image Compression using Genetic Algorithm, IEEE Trans Image Process. pp. 586-93, 1998.

17. Mumtaz A. and Majid A., Genetic Algorithms and its application to Image Fusion, International Conference On Emerging Technologies ICET, 2008.

18. Patnaik D., Biomedical Image Fusion using Wavelet Transforms and Neural Network, IEEE International Conference on Industrial Technology, pp. 1189 – 1194, 2006.

19. Pattichis C. S. and Pattichis M. S., Medical Imaging Fusion Applications— An Overview, In Conf. Rec. Asilomar Conf. Signals, Systems Computers, Vol. 2, pp. 1263–1267, 2001.

20. Prokoski F., History, Current Status, and Future of Infrared Identification, Computer Vision Beyond the Visible Spectrum: Methods and Applications, Proceedings IEEE Workshop, pp. 5-14, 2000.

21. Rahman Z., Jobson D. and Woodell G. A., Multiscale Retinex For Color Image Enhancement, In Proceedings of the IEEE International Conference On Image Processing, 1996.

22. Rahman Z., The Lectures Notes of Image Processing, Old Dominion University, 2009.

23. Rahman Z., Woodell G. A. and Jobson D., A Comparison of The Multiscale Retinex with other Image Enhancement Techniques, In Proceedings Of The IS&T 50th Anniversary Conference, pp. 426-431, 1997b.

24. Schreiber W. F., Image processing for quality improvement, Proceedings of the IEEE, Vol. 66, pp. 1640–1651, 1978.

25. Socolinsky D. A. and Selinger A., A Comparative Analysis of Face Recognition Performance with Visible and Thermal Infrared Imagery, IEEE International Conference of Pattern Recognition, Vol. 4, pp. 217 –222, 2002.

26. Srinivas M. and Patnaik L. M., Genetic Algorithms: a Survey, pp. 17 – 26, Vol. 27, Jun 1994.

27. Tao L., An Adaptive And Integrated Neighborhood Dependent Approach For Nonlinear Enhancement Of Color Images, SPIE Journal of Electronic Imaging, pp. 1.1-1.14, 2005.

28. Tao L., Tompkins R. C., and Asari K. V., An Illuminance Reflectance Model For Nonlinear Enhancement Of Video Stream For Homeland Security Applications, IEEE International Workshop on Applied Imagery and Pattern Recognition, AIPR, October 19 - 21, 2005.

29. Wang A., Sun H. and Guan Y., The Application of Wavelet Transform to Multi-Modality Medical Image Fusion, Networking, Sensing and Control, ICNSC Proceedings Of The 2006 IEEE International Conference, pp. 270-274, 2006.

30. Wang H., Multisensory Image Fusion by using Discrete Multiwavelet Transform, The Third International Conference on Machine Learning and Cybernetics, Shanghai, 26-29 August 2004.

31. Wilson T. A., and Rogers S. K., Perceptual-Based Image Fusion for Hyperspectral Data, IEEE Trans. Ge. Remote Sensing, Vol. 35, pp. 1007–1017, July 1997.

32. Yu L., Yung T., Chan K., Ho Y. and Ping Chu Y., Image Hiding with an improved Genetic Algorithm and an Optimal Pixel Adjustment Process, Eighth International Conference On Intelligent Systems Design And Applications, 2008.

33. Zhang J., Feng X., Song B., Li M. and Lu Y., Multi-Focus Image Fusion using Quality Assessment of Spatial Domain And Genetic Algorithm, Human System Interactions, pp. 71 – 75, 25-27 May 2008.

34. Zitova B. and Flusser J., Image Registration Methods: A Survey, Image and Vision Computing 21, pp. 977–1000, 2003.

Chapter 6

OPTIMAL FEATURE GENERATION WITH GENETIC ALGORITHMS AND FLDR IN A RESTRICTEDVOCABULARY SPEECH RECOGNITION SYSTEM

Julio César Martínez-Romo[1], Francisco Javier Luna-Rosas[2], Miguel Mora-González[3], Carlos Alejandro de Luna-Ortega[4] and Valentín López-Rivas[5]

[1,2,5]Instituto Tecnológico de Aguascalientes, Mexico

[3]Universidad de Guadalajara, Centro Universitario de los Lagos, Mexico

[4]Universidad Politécnica de Aguascalientes, Mexico

INTRODUCTION

In every pattern recognition problem there exist the need for variable and feature selection..and, in many cases, feature generation. In pattern recognition, the term variable is usually ..understood as the raw measurements or raw values taken from the subjects to be classified, ..while the term feature is used to refer to the result of the transformations applied to the ..variables in order to transform them into another domain or space, in which a bigger ..discriminant capability of the new calculated features is expected; a very popular cases of ..feature generation are the use of principal component analysis (PCA), in which the variables ..are projected into a lower dimensional space in which the new features can be used to ..visualize the underlying class distributions in the original data [1], or the Fourier Transform, ..in which a few of its coefficients can represent new features [2], [3]. Sometimes, the ..literature does not make any distinction between variables and features, using them ..indistinctly [4], [5]. ..Although many variables and features can be obtained for classification, not all of them ..posse discriminant capabilities; moreover, some of them could cause confusion to a ..classifier. That is the reason why the designer of the classification system will require to ..refine his choice of variables and features. Several specific techniques for such a purpose are ..available [1], and some of them will be reviewed later on in this chapter. ..Optimal feature generation is the generation of the features under some optimality criterion,

..usually embodied by a cost function to search the solutions' space of the problem at hand ..and providing the best option to the classification problem. Examples of techniques like ..these are the genetic algorithms [6] and the simulated annealing [1]. In particular, genetic ..algorithms are used in this work. ..Speech recognition has been a topic of high interest in the research arena of the pattern ..recognition community since the beginnings of the current computation age [7], [8]; it is due,partly, to the fact that it is capable of enabling many practical applications in artificial intelligence, such as natural language understanding [9], man-machine interfaces, help for the impaired, and others; on the other hand, it is an intriguing intellectual challenge in which new mathematical methods for feature generation and new and more sophisticated classifiers appear nearly every year [10], [11]. Practical problems that arise in the implementation of speech recognition algorithms include real-time requirements, to lower the computational complexity of the algorithms, and noise cancelation in general or specific environments [12]. Speech recognition can be user or not-user dependant. A specific case of speech recognition is word recognition, aimed at recognizing isolated words from a continuous speech signal; it find applications in system commanding as in wheelchairs, TV sets, industrial machinery, computers, cell phones, toys, and many others. A particularity of this specific speech processing niche is that usually the vocabulary is comprised of a relatively low amount of words; for instance, see [13] and [14]. In this chapter we present an innovative method for the restricted-vocabulary speech recognition problem in which a genetic algorithm is used to optimally generate the design parameters of a set of bank filters by searching in the frequency domain for a specific set of sub-bands and using the Fisher's linear discriminant ratio as the class separability criterion in the features space. In this way we use genetic algorithms to create optimum feature spaces in which the patterns from N classes will be distributed in distant and compact clusters. In our context, each class $\{\omega_0, \omega_1, \omega_2,..., \omega_{N-1}\}$ represents one word of the lexicon. Another important part of this work is that the algorithm is required to run in real time on dedicated hardware, not necessarily a personal computer or similar platform, so the algorithm developed should has low computational requirements. This chapter is organized as follows: the section 2 will present the main ideas behind the concepts of variable and feature selection; section 3 presents an overview of the most representative speech recognition methods. The section 4 is devoted to explain some of the mathematical foundations of our method, including the Fourier Transform, the Fisher's linear discriminant ratio and the Parseval's theorem. Section 5 shows our algorithmic foundations, namely the genetic algorithms and the back propagation neural networks, a powerful classifier used here for performance comparison purposes. The implementation of our speech recognition approach is depicted in section 6 and, finally, the conclusions and the future work are drawn in section 7.

OPTIMAL VARIABLE AND FEATURE SELECTION

Feature selection refers to the problem of selecting features that are most predictive of a given outcome. Optimal feature generation, however, refers to the derivation of features from input variables that are optimal in terms of class separability in the feature space. Optimal feature generation is of particular relevance to pattern recognition problems because it is the basis for achieving high correct classification rates: the better the discriminant features are represented, the better the classifier will categorize new incoming patterns. Feature generation is responsible for the way the patterns lay in the features space, therefore, shaping the decision boundary of every pattern recognition problem; linear as well as non-linear classifiers can be beneficiaries of well-shaped feature spaces.

The recent apparition of new and robust classifiers such as support vector machines (SVM), optimum margin classifiers and relevance vector machines [4], and other robust kernel classifiers seems to demonstrate that the new developments are directed towards classifiers which, although powerful, must be preceded by reliable feature generation techniques. In some cases, the classifiers use a filter that consists of a stage of feature selection, like in the Recursive Feature Elimination Support Vector Machine [15], which eliminates features in a recursive manner, similar to the backward/forward variable selection methods [1].

Methods for variable and feature selection and generation

The methods for variable and feature selection are based on two approaches: the first is to consider the features as scalars -scalar feature selection-, and the other is to consider the features as vectors –feature vector selection-. In both approaches a class separability measurement criteria must be adopted; some criteria include the receiver operating curve (ROC), the Fisher Discriminant Ratio (FDR) or the one-dimensional divergence [1]. The goal is to select a subset of k from a total of K variables or features. In the sequel, the term features is used to represent variables and features.

Scalar feature selection

The first step is to choose a class separability measuring criterion, C(K). The value of the criterion C(K) is computed for each of the available features, then the features are ranked in descending order of the values of C(K). The k features corresponding to the k best C(K) values are selected to form the feature vector. This approach is simple but it does not take into consideration existing correlations between features.

Vector feature selection

The scalar feature selection may not be effective with features with high mutual correlation; another disadvantage is that if one wishes to verify all possible combinations of the features –in the spirit of optimality- then it is evident that the computational burden is a major limiting factor. In order to reduce the complexity some suboptimal procedures have been suggested [1]:

Sequential Backward Selection" . The following steps comprise this method:

- Select a class separability criterion, and compute its value for the feature vector of all the features.
- Eliminate one feature and for each possible combination of the remaining features recalculate the corresponding criterion value. Select the combination with the best value.
- From the selected K-1 feature vector eliminate one feature and for each of the resulting combinations compute the criterion value and select the one with the best value.
- Continue until the feature vector consists of only the k features, where k is the predefined size.

The number of computations can be calculated from: $1+1/2 ((K+1)K - k(k+1))$.

a. Compute the criterion value for each individual feature; select the feature with the "best" value,
b. From all possible two-dimensional vectors that contains the winner from the previous step. Compute the criterion value for each of them and select the best one.

Floating search methods

The methods explained suffer from the nesting effect, which means that once a feature (or variable) has been discarded it can't be reconsidered again. Or, on the other hand, once a feature (or variable) was chosen, it can't be discarded. To overcome these problems, a technique known as floating search method, was introduced by Pudin and others in 1994 [1], allowing the features to enter and leave the set of the k chosen features. There are two ways to implement this technique: one springs from the forward selection and the other from de backward selection rationale. A three steps procedure is used, namely inclusion, test, and exclusion. Details of the implementation can be found in [1], [16].

Some trends in feature selection

Recent work in feature selection are, for instance, the one of Somol et al., [17], where besides of optimally selecting a subset of features, the size of the subset is also optimally selected. Sun and others [18] faced the problem of feature selection in conditions of a huge number or irrelevant features, using machine learning and numerical analysis methods without making any assumptions about the underlying data distributions. In other works, the a feature selection technique is accompanied by instance selection; instance selection refers to the "orthogonal version of the problem of feature selection" [19], involving the discovery of a subset of instances that will provide the classifier with a better predictive accuracy than using the entire set of instances in each class.

Optimal feature generation

As can be seen from section 2.1, the class separability measuring criterion in feature selection is used just to measure the effectiveness of the k features chosen out of a total of K features, with independence of how the features were generated. The topic of optimal feature generation refers to involving the class separability criterion as an integral part of the feature generation process itself. The task can be expressed as: If x is an m-dimensional vector of measurement samples, transform it into another l-dimensional vector andso that some class separability criterion is optimized. Consider, to this end, the linear transformation y=ATx. By now, it will suffice to note the difference between feature selection and feature generation.

SPEECH RECOGNITION

For speech processing, the electrical signal obtained from an electromechanoacoustic transducer is digitized and quantized at a fixed rate (the sampling frequency, F_s), and subsequently segmented into small frames of a typical duration of 10 milliseconds. Regarding to section 2, the raw digitized values of the voice signal will be considered here as the input variable; the mathematical transformations that will be applied to this variable will produce the features. Two important and widely used techniques for speech recognition will be presented in this section due to its relevance to this field. Linear Predictive Coding, or LPC, is a predictive technique in which a linear combination of some K coefficients and the last K samples from the signal will predict the value of the next one; the K coefficients will represent the distinctive features. The following section will explain the LPC method in detail.

Linear Predictive Coding (LPC)

LPC is one of the most advanced analytical techniques used in the estimation of patterns, based on the idea that the present sample can be predicted from a linear combination of some past samples, generating a spectral description based on short segments of signal considering a signal s[n] to be a response of an all-pole filter excitation u[n].

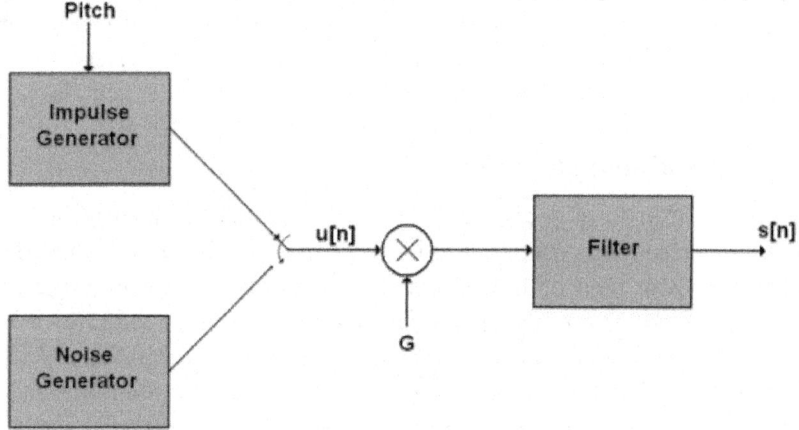

Figure. 1: LPC model of speech

Figure 1 shows the model the LPC is based on, considering that the excitation u[n] is the pattern waiting to be recognized. The transfer function of the filter is described as [3]:

$$H(z) = \frac{S(z)}{U(Z)} = \frac{G}{1 - \sum_{k=1}^{p} a_k z^{-k}} = \frac{G}{A(z)},$$

$$(1)$$

where G is a gain parameter, a_k are the coefficients of filter and p determines the order of the filter. In Figureure 1, the samples s[n] are related to the excitation u[n] by the equation:

$$s[n] = \sum_{k=1}^{p} a_k s[n-k] + Gu[n],$$

$$(2)$$

Considering that the linear combination of past samples is calculated by using an estimator $\tilde{s}[n]$ which is denoted by

$$\tilde{s}[n] = \sum_{k=1}^{p} a_k s[n-k],$$

(3)

the error in the prediction is determined by the lack of accuracy with respect to s[n], which is defined as [20]:

$$e[n] = s[n] - \tilde{s}[n] = s[n] - \sum_{k=1}^{p} a_k s[n-k],$$

(4)

$$e[z] = s[z]\left(1 - \sum_{k=1}^{p} a_k z^{-k}\right).$$

(5)

from equation (5), it is possible to recognize that the sequence of prediction of the error has in its components a FIR-type filter system which is defined by:

$$A(z) = 1 - \sum_{k=1}^{p} a_k z^{-k} = \frac{E(z)}{S(z)},$$

(6)

equations (2) and (4) show that e[n]=Gu[n]. The estimation of the prediction coefficients is obtained by minimizing the error in the prediction. Where e[n]^2 denotes the square error of the prediction and E is the total error over a time interval (m). The prediction error in a short time segment is defined as:

$$E = \sum_{m} e[m]^2 = \sum_{m} (s[m] - \sum_{k=1}^{p} a_k s[m-k])^2,$$

(7)

the coefficients $\{a_k\}$ minimize the prediction of error E on the fragment obtained by the partial derivatives of E with respect to such coefficients; this means that:

$$\frac{\partial E}{\partial a_i} = 0, \quad 1 \le i \le p.$$

(8)

Through equations (7) and (8) the final equation is obtained:

$$\sum_{k=1}^{p} a_k \sum_{n} s[n-k]s[n-i] = -\sum_{n} s[n]s[n-i], \qquad 1 \le i \le p,$$

(9)

this equation is written in terms of least squares and is known as a normal equation. For any definitions of the signal s[n], equation (9) forms a set of p equations with p unknowns that must be solved for coefficients $\{a_k\}$, trying to reduce the error E of equation (7). The minimum total squared error, denoted by E_p, is obtained by expanding equation (7) and substituting the result in equation (9), this is:

$$Ep = \sum_n s^2[n] + \sum_{k=1}^{p} a_k \sum_n s[n]s[n-k],$$

(10)

using the autocorrelation method to solve it [8].

For application, it is assumed that the error of equation (7) is minimized for infinite duration defined as $-\infty<n$ thus equations (9) and (10) are simplified as:

$$\sum_{k=1}^{p} a_k R(i-k) = -R(i), \qquad\qquad 1 \le i \le p,$$

(11)

$$Ep = R(0) + \sum_{k=1}^{p} a_k R(k),$$

(12)

Where

$$R(i) = \sum_{n=-\infty}^{\infty} s[n]s[n+i],$$

(13)

which is the autocorrelation function of the signal s[n], with R(i) as an even function. The coefficients R(i-k) generate auto-correlation matrix, which is a symmetric Toeplitz matrix; ie, all elements in each diagonal are equal. For practical purposes, the signal s[n] is analyzed in a finite interval. One popular method of approach this is by multiplying the signal s'[n] times a window function w[n] in order to obtain an s'[n] signal:

$$s'[n] = \begin{cases} s[n]w[n], & 0 \le n \le N-1 \\ 0, & \text{otherwise} \end{cases}$$

(14)

Using equation 14, the auto-correlation function is given by:

$$R(i) = \sum_{n=0}^{N-1-i} s'[n]s'[n+i], \qquad i \ge 0.$$

(15)

One of the most common ways to find the coefficients $\{a_k\}$, is by computational methods, where the equation (11) is expanded to a matrix with the form:

$$\begin{bmatrix} R_0 & R_1 & R_2 & \cdots & R_{p-1} \\ R_1 & R_0 & R_1 & \cdots & R_{p-2} \\ R_2 & R_1 & R_0 & \cdots & R_{p-3} \\ \vdots & \vdots & \vdots & & \vdots \\ R_{p-1} & R_{p-2} & R_{p-3} & \cdots & R_0 \end{bmatrix} \begin{bmatrix} a_1 \\ a_2 \\ a_3 \\ \vdots \\ a_p \end{bmatrix} = \begin{bmatrix} R_1 \\ R_2 \\ R_3 \\ \vdots \\ R_p \end{bmatrix},$$

and it is necessary to use an algorithm to find these coefficients; one of the most commonly used, is the Levinson-Durbin one, which is described below [21]:

$$E^0 = R(0)$$
$$a_0 = 1$$
$$\text{for } i = 1, 2, ..., p$$

$$k_i = \frac{\left(R(i) - \sum_{j=1}^{i-1} a_j^{(i-1)} R(i-j) \right)}{E^{(i-1)}}$$

$$a_i^{(i)} = k_i$$

$$\text{if } i > 1 \text{ then for } j = 1, 2, ..., i-1$$
$$a_j^{(i)} = a_j^{(i-1)} - k_i a_{i-j}^{(i-1)}$$

end

$$E^{(i)} = (1 - k_i^2) E^{(i-1)}$$

end

$$a_j = a_j^{(p)} \quad j = 1, 2, ..., p.$$

An important feature of this algorithm is that, when making the recursion, an estimation of the half-quadratic prediction error must be made. This prediction satisfies the system function given in equation (17), which corresponds to the term A (z) of equation (1); namely:

$$A^{(i)}(z) = A^{(i-1)}(z) - k_i z^{-i} A^{(i-1)} z^{-1}, \tag{17}$$

where the fundamental part for the characterization of the signal in coefficients of prediction, is met by establishing an adequate number of coefficients p, according to the sampling frequency (f_s) and based on the resonance in kHz [3] which is:

$$p = 4 + \frac{f_s}{1000} \tag{18}$$

where the optimal number of LPC coefficients is the one that represents the lowest mean square error possible. Figure 2 shows the calculation of LPC in a voice signal with 8 kHz sampling rate and the effect of varying the number of coefficients in a segment (frame).

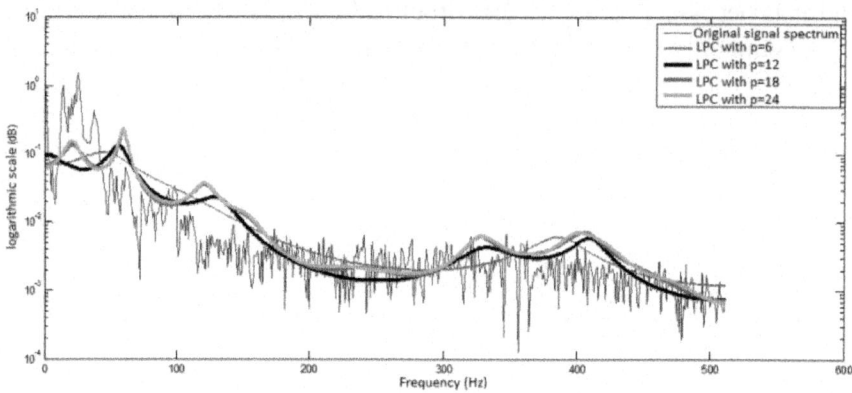

Figure. 2: Comparison of the original signal spectrum and LPC envelope with different numbers of coefficients

Dynamic Time Warping (DTW)

Another commonly used technique used in speech recognition is the dynamic time warping. Is presented here, again, for it relevance to the this field. Dynamic time warping is a technique widely used in pattern recognition, particularly oriented to temporal distortions between vectors, such as the time of writing, speed of video camera, the omission of a letter, etc. These temporal variations are not proportional and vary accordingly to each person, object or event, and those situations are not repetitive in any aspects. DTW uses dynamic programming to find similarities and differences between two or more vectors.

This method considers two sequences representing feature vectors defined by a(i), i=1,2,...,I and b(j), j=1,2,...,J, where, in general, the number of elements differs in each vector (I≠J). The aim of DTW is to find an appropriate distance between the two sequences and in a two-dimensional plane, where each sequence represents one axis, and each point corresponds to the local relationship between two sequences. The nodes (i,j) of the plane are associated with a cost that is defined by the function $d(c)=d(i,j)=|a(i)-b(j)|$, which represents the distance between the elements a(i) and b(j).

The collection of points begins in the starting point (i_0,j_0) and finishes in the (i_k,j_k) nodes, and it being an ordered pair of size k, where k is the number of nodes along the way. Every path established with the points is associated with a total cost D and defined by

$$D = \sum_{k=0}^{K-1} d(i_k, j_k),$$

$$(19)$$

and the distance between the two sequences is defined as the minimum value D of all the possible paths

$$D(a,b) = \min_k(D).$$

$$(20)$$

There are normalization and temporary limitations in the search for the minimum distance between patterns to compare [22], [1]. These limitations are: endpoint, monotonicity conditions, local continuity, global path and slope weight. The final point is bounded by the size of windowing and performed in each pattern, at most cases is empirical and defined to extremes, that is:

$$i(1) = 1, j(1) = 1$$
$$i(K) = I, j(K) = J.$$

$$(21)$$

Figure 3 shows an example in which it is only partially considered one of the sequences, a situation that is not allowed to search the minimum cost. The monotonicity conditions try to maintain the temporal order of the normalization of the time, and avoid negative slopes,

$$i_{k+1} \geq i_k$$

$$(22)$$

And

$$j_{k+1} \geq j_k.$$

$$(23)$$

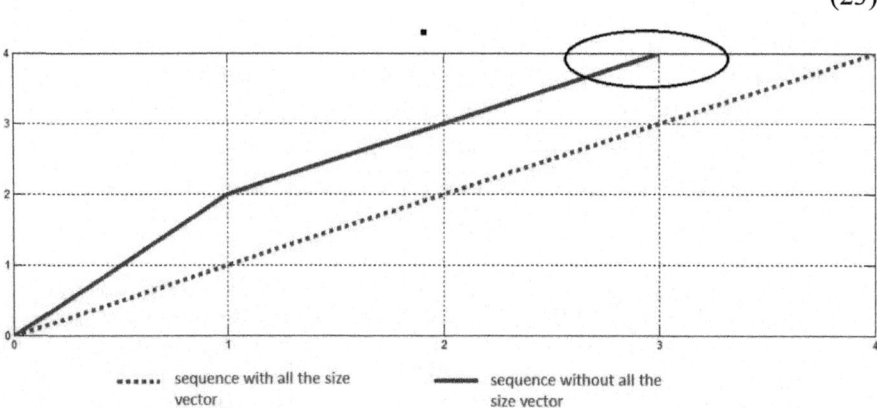

Figure. 3: Example of a sequence that violates the rule of the endpoint.

Figure 4 shows an example without monotonicity paths, which are not allowed to find the optimal path

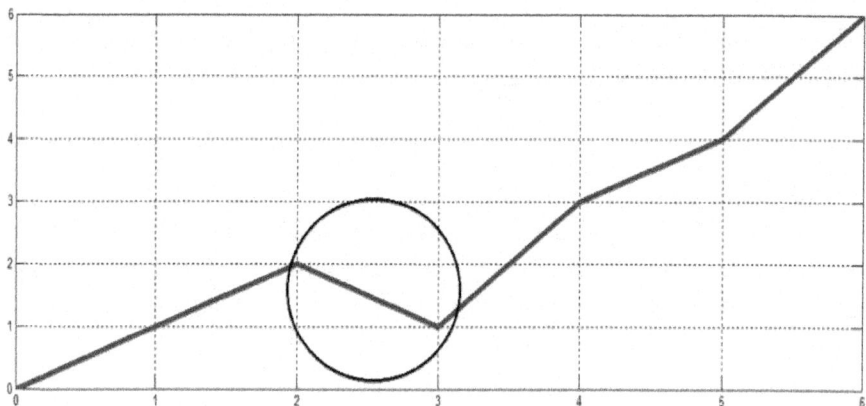

Figure. 4: Without monotonicity path example.

The continuity conditions

$$i(K) - i(k-1) \leq 1 \quad \text{And} \quad j(K) - j(K-1) \leq 1,$$

$$(24)$$

are defined by maintaining the relationship between two consecutive points of the form:

$$c(K-1) = \begin{cases} (i(k), j(k)-1), \\ (i(k)-1, j(k)-1), \\ (i(k)-1, j(k)), \end{cases}$$

$$(25)$$

Global limitations define a region of nodes where the optimal path is found, and is based on a parallelogram that offers a feasible region [7], thereby avoiding unnecessary regions involved in processing. Figure 5 shows the values of the key points of the parallelogram.

The optimal path layout defined a measure of dissimilarity between the two sequences of features, whose general form is

$$D(a,b) = \min_{F} \left[\frac{\sum_{k=1}^{K} d(c(k)) \cdot w(k)}{\sum_{k=1}^{K} w(k)} \right],$$

$$(26)$$

where d(c(k)) and w(k) are the local distance between the windows i(k) of the reference vector and j(k) of the recognize vector, and a weighting function in k to maintain a flexible way and improve alignment, respectively. The simplified computational algorithm for calculating the distance of DTW is shown below [1]

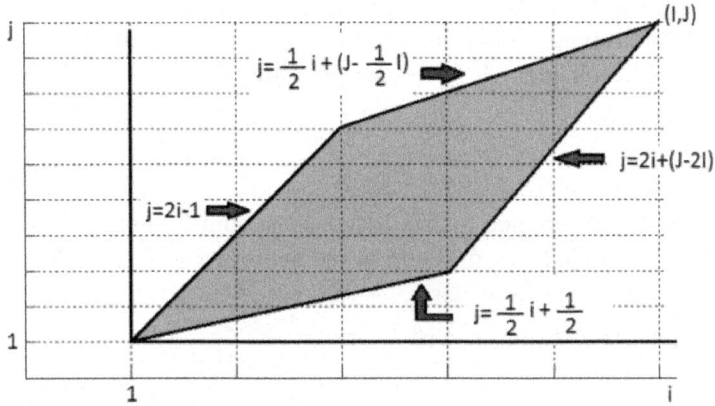

Figure. 5: Global region and determination of slopes.

$$D(0,0) = 0$$
for $i = 1 : I$
$$\qquad D(i,0) = D(i-1,0) + 1$$
end
for $j = 1 : J$
$$\qquad D(0,j) = D(0,j-1) + 1$$
end
for $i = 1 : I$
$$\qquad \text{for } j = 1 : J$$
$$\qquad\qquad c1 = D(i-1,j-1) + d(i,j|i-1,j-1)$$
$$\qquad\qquad c2 = D(i-1,j) + 1$$
$$\qquad\qquad c3 = D(i,j-1) + 1$$
$$\qquad\qquad D(i,j) = \min(c1,c2,c3)$$
$$\qquad \text{end}$$
end
$$D(a,b) = D(I,J)$$

MATHEMATICAL FOUNDATIONS

Fisher's Linear Discriminant Ratio FLDR

Fisher's Linear Discriminant Ratio, is used as an optimization criterion in several research fields, including speech recognition, handwriting recognition, and others [1]. Consider the following definitions: Within-class scatter matrix.

$$S_w = \sum_{i=1}^{M} P_i S_i$$

(27)

where S_i is the covariance matrix for class ω_i, and P_i the a priori probability of class ω_i. Trace$\{S_w\}$ is a measure of the average variance of the features, or descriptive elements of the class.

Between-class scatter matrix

$$S_b = \sum_{i=1}^{M} P_i(\mu_i - \mu_0)(\mu_i - \mu_0)^T$$

(28)

where μ_0 is the global mean vector

$$\mu_0 = \sum_{i}^{M} P_i \mu_i$$

(29)

Trace$\{S_b\}$ is a measure of the average distance of the mean of each individual class from the respective global value.

Mixture scatter matrix

$$S_m = E\left[(\mu_i - \mu_0)(\mu_i - \mu_0)^T\right]$$

(30)

S_m is the covariance matrix of the feature vector with respect to the global mean, and E[.] is the mathematical operator of the expected mean value. Based on the just given definitions, the following criteria can be expressed:

$$J_1 = \frac{trace\{S_m\}}{trace\{S_w\}}$$

$$J_2 = \frac{|S_b|}{|S_w|} = |S_w^{-1}S_b|$$

(31)

It can be shown that J1 and J2 take large values when the samples in the l-dimensional are well clustered around their mean, within each class, and the clusters of different classes are well separated. Criteria in equation (31) can be used to guide an optimization process, since they measure the goodness of data clustered; the data to be clustered could be the set of features representative of the items of a class. Trace$\{S_b\}$ is a measure of the average distance of the mean of each individual class from the respective global value.

Figureure 6 shows an example in which the FLDR is evaluated using equation (31); FLDR and the respective values are displayed. Notice that the more the blue clusters are separated from the red cluster, the bigger FLDR value is.

Fldr Example: The More Separate The Clusters Are The Bigger J Is

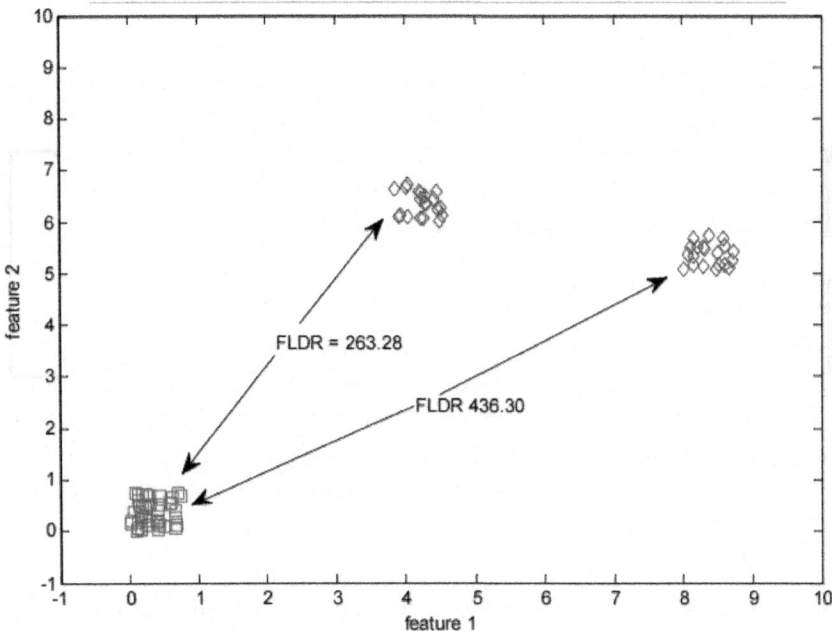

Figure. 6: Example of the FLDR values for two clusters.

Parseval's theorem and the Fourier

Transform Parseval's theorem states, in an elegant manner, that the energy of a discrete signal in the time domain can be calculated in the frequency domain by a simple relation [2], [3]:

$$\sum_{i=0}^{N-1} x[i]^2 = \frac{2}{N} \sum_{k=0}^{N/2} |X[k]|^2$$

(32)

where N is the number of samples of the discrete signal,

x[i] is the i-th sample of the discrete signal,

X[k] is the k-th sample of the Fourier transform of

x[i]. For a discrete signal

x[i], the Fourier transform can be computed using the well known

Discrete Fourier Transform via the efficient algorithm for its implementation, the FFT [2], [3]:

$$X[k] = \sum_{n=0}^{N-1} x[n]e^{-2i\pi k\frac{n}{N}} \quad k=0, 1,...,N-1$$

(33)

The implication of the Parseval's theorem is that an algorithm can search for specific energetic properties of a signal in the frequency domain off-line, and then use the information obtained off-line to conFigureure a bank of digital filters to look for the same energetic properties in the time domain on-line, in real time. The link between both domains is the energetic content of the signal.

ALGORITHMIC FOUNDATIONS

This section is devoted to describe two important Figureures in pattern recognition: backpropagation neural networks BPNN and genetic algorithms GA. The BPNN is used as a reference classifier to compare the performance of the approach presented here to the word recognition problem. The GA is an integral part of the generation of the features in the proposed technique

Learning paradigms

There are several major paradigms, or approaches, to machine learning. These include supervised, unsupervised, and reinforcement learning. In addition, many researchers and application developers combine two o more of these learning approaches into one system [23]. Supervised learning is the most common form of learning and is sometimes called programming by example. The learning system is trained by showing it examples of the problem state or attributes along with the desired output or action. The learning system make a prediction based on the inputs and if the output differs from the desired output, then the system is adjusted or adapted to produce the correct output. This process is repeated over and over until the system learns to make accurate classifications or predictions. Historical data from databases, sensor logs, or trace logs is often used as the training or example data. Unsupervised learning is used when the learning system needs to recognize similarities between inputs or to identify features in the input data. The data is presented to the system, and it adapts so that it partitions the data into groups. The clustering or segmenting process continues until the system places the same data into the same group on successive passes over the data. An unsupervised learning algorithm performs a type of feature detection where important common attributes in the data are extracted. Reinforcement learning is a type of supervised learning used when explicit input/output pairs of training data are not available. It can be used in cases where there is a sequence of inputs and the desired output is only known after the specific sequence occurs. This process of identifying the relationship between a series of input values and a later output value is called temporal credit assignment. Because we provide less specific error information, reinforcement learning usually takes longer than supervised learning and is less efficient. However, in many situations, having exact prior information about the desired outcome is not possible. In many ways, reinforcement learning is the most realistic form of learning. Another important distinction in learning systems is whether the learning is done on-line or off-line. On-line learning means that the system is sent out to perform its tasks and that it can learn or adapt after each transaction is processed. On-line learning is like on the job training and places severe requirements on the learning algorithms. It must be very fast and very stable. Off-line learning, on the other hand, is more like a business seminar. You take your salespeople off the floor and place them in an environment where they can focus on improving their skills without distractions. After a suitable training period, they are sent out to apply their new found knowledge and skills. In an intelligent system context, this means that we would gather data from situations that the systems have experienced. We could then augment this data with information about the desired system response to build a training data set. Once we have this database we can use it to modify the behavior of our system.

BACKPROPAGATION NEURAL NETWORKS

Backpropagation is the most popular neural network architecture for supervised learning. It features a feed-forward connection topology, meaning that data flow through the network in a single direction, and uses a technique called the backward propagation of errors to adjust the connection weights Rumelhart, Hinton, and Williams 1986 in [23]. In addition to a layer of input and output units, a back-propagation network can have one or more layers of hidden units, which receive inputs only from other units, and not from the external environment. A back propagation network with a single hidden layer or processing units can learn to model any continuous function when given enough units in the hidden layer. The primary applications of back propagation networks are for prediction and classification. Figureure 7 shows the diagram of a back propagation neural network and illustrates the three major steps in the training process.

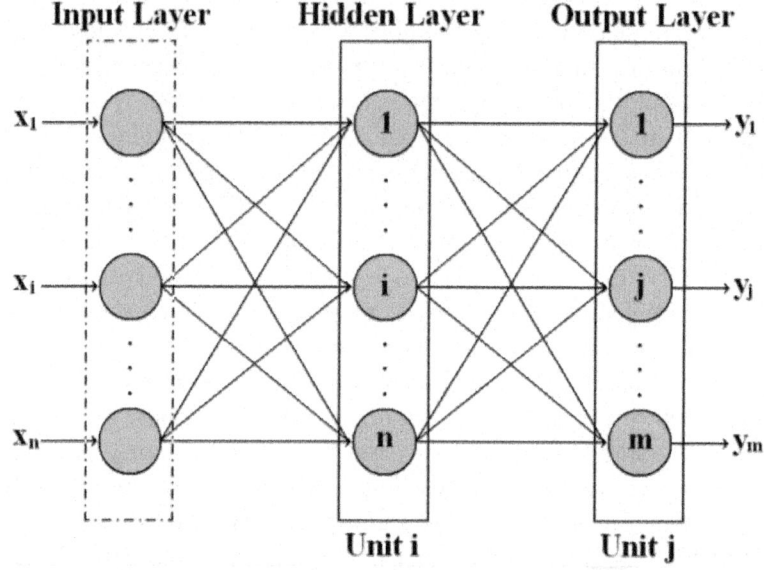

Figure. 7: Topology of a backpropagation neural network.

First, input data is presented to the units of the input layer on the left, and it flows through the network until it reaches the network output units on the right. This is called the forward pass.

Second, the activations or values of the output units represent the actual or predicted output of the network, because this is supervised learning.

Third, the difference between the desired and the actual output is computed, producing the network error. This error term is then passed backwards through the network to adjust the connection weights. Each network input unit takes a single numeric value, x_i , which is usually scaled or normalized to a value between 0.0 and 1.0. This value becomes the input unit activation. Next, we need to propagate the data forward, through the neural network. For each unit in the hidden layer, we compute the sum of the products of the input unit activations and the weights connecting those input layer units to the hidden layer. This sum is the inner product (also called the dot or scalar product) of the input vector and the weights in the hidden unit. Once this sum is computed, we add a threshold value and then pass this sum through a nonlinear activation function, f , producing the unit activation y_i . The formula for computing the activation of any unit in a hidden or output layer in the network is

$$y_i = f\left(sum_j = \sum x_i w_{ij} + \theta_i\right)$$

(34)

where i ranges over all the units leading into the j-th unit, and the activation function is

$$f\left(sum_j\right) = \frac{1}{1 + e^{-sum_j}}$$

(35)

As mentioned earlier, we use the S-shape sigmoid or logistic function for f. The formula for calculating the changes of the weights is

$$\Delta w_{ij} = \eta \delta_j y_i$$

(36)

where w_{ij} is the weight connecting unit i to unit j, η is the learn rate parameter, j δ is the error signal for that unit, and y_i is the output or activation value of unit i. For units in the output layer, the error signal is the difference between the target output j t and the actual output y_i multiplied by the derivative of the logistic activation function.

$$\delta_j = \left(t_j - y_j\right) f_j'\left(sum_j\right) = \left(t_j - y_j\right) y_j \left(1 - y_j\right)$$

(37)

For each unit in the hidden layer, the error signal is the derivative of the activation function multiplied by the sum of the products of the outgoing connection weights and their corresponding error signals. So for the hidden unit j.

$$\delta_j = f_j'\left(sum_j\right) \sum \delta_k w_{jk}$$

(38)

where k ranges over the indices of the units receiving j-th unit's output signal. A common modification of the weight update rule is the use of a momentum term α, to cut down on oscillation of the weight change becomes a combination of the current weight change, computed as before, plus some fraction (α ranges from 0 to 1) of the previous weight change. This complicates the implementation because we now have to store the weight changes from the prior step.

$$\Delta w_{ij}\left(n + 1\right) = \eta \delta_j y_i + \alpha \Delta w_{ij}\left(n\right)$$

(39)

The mathematical basis for backward propagation is described in detail in [23]. When the weight changes are summed up (or batched) over an entire presentation of the training set, the error minimization function performed is called gradient descent.

Genetic Algorithms In this section a brief description of a simple genetic algorithm is given. Genetic algorithms are based on concepts and methods observed in nature for the evolution of the species. Genetic algorithms were brought to the artificial intelligence arena by Goldberg [6], [24]. They apply certain operators to a population of solutions of the problem to be solved, in a such a way that the new population is improved compared to the previous one according to a certain criterion function J [5], [1], [6], [24]. Repetition of this procedure for a preselected number of iterations will produce a last generation whose best solution is the optimal solution to the problem. The solutions of the problem to be solved are coded in the chromosome and the following operations are applied to the coded versions of the solutions, in this order: Reproduction. Ensures that, in probability, the better a solution in the current population is, the more replicates it has in the next population, Crossover. Selects pair of solutions randomly, splits them in a random position, and exchanges their second parts. Mutation. Selects randomly an element of a solution and alters it with some probability. It helps to move away from local minima. Besides the coding of the solutions, some parameters must be set up: N, number of solutions in a population. Fixed or varied. p, probability with which two solutions are selected for crossover. m, probability with which an element of a solution is mutated. The performance of the GA depends greatly on these parameters, as well as on the coding of the solutions in the chromosome. The solutions can be coded in some of the following formats: Binary. Bit strings represent the solution(s) of the problem. For instance, a chromosome could represent a series of integer indexes to address a database, or the value of a variable(s) that must be integer, or each bit could represent the state (present-absent) of a part of an architecture that is being optimized, and so on. Real valued. The bit strings represent the value of a real valued variable, in fixed of floating point. The aspect of one chromosome could be like this: C = {100101010101010101}; the interpretation will vary in accordance with the coding scheme selected to represent the knowledge domain of the problem. For instance, it might represent a set of six indices of three bits each one; or it could have a meaning with all the bits together, representing an 18 bit code.

The primary reason of the success of genetic algorithms is its wide applicability, easy use and global perspective [6], [24], [25]. The next is the listing of a simple genetic algorithm.

1.	**Procedure** (Genetic_Algorithm)
2.	M = Population size. (*# Of possible solutions at any instance.*)
3.	N_g = Number of generations. (*# Of iterations.*)
4.	N_o = Number of offsprings. (*# To be generated by crossover.*)
5.	P_μ = Mutation probability. (*# Also called mutation rate M_r.*)
6.	P ← Ξ (M)
7.	**For** j = 1 to M
8.	Evaluate $f(p[i])$
9.	**EndFor**
10.	**For** i = 1 to N_g
11.	**For** (j=1 to N_o)
12.	(x,y) ←Ø(p) (*Select two parents x and andfrom current population*)
13.	Offspring[j] ← X(x,y) (*Generate offsprings by crossover of parents x and y*)
14.	Evaluate f(offspring[j]) (*Evaluate fitness of each offsprings*)
15.	**EndFor**
16.	**For** j = 1 to N_o (*With probability p_μ apply mutation*)
17.	Mutated[j] ← μ(y)
18.	Evaluate f(mutated[j])
19.	**EndFor**
20.	p ← Selected(p, iffsprings) (*Select best M solutions from parents & offsprings. *)
21.	**EndFor**
22.	Return highest scoring configuration in p.
23.	**End**

The genetic algorithms find application in the field of speech processing via the solution to the problem of variable and feature selection [11], [14], [26], [27], [28], [29], [30].

RESTRICTED-VOCABULARY SPEECH RECOGNITION SYSTEM

The expression restricted-vocabulary speech recognition refers to the recognition of repetitions of spoken words that belongs to a limited set of words within a semantic field. This means that the words have connected meanings, for instance, the digits = {0,1,2,..., 9} or the days of the week={Saturday, Monday,..., Friday}. The applications of the recognition of limited size word-sets include voice-commanded systems, spoken entry and search for computer databases in warehouse systems, voice-assisted telephone dialing, man-machine interfaces, and others. The advantage of a system developed for a specific semantic field is that it can be built to be much more accurate than those constructed for the general speech recognition, also requiring less extensive training sets. Restricted-vocabulary speech recognition is also an important research topic because of the intricacies involved in the underlying pattern recognition problem: variable selection, feature generation/selection and classifier selection. Variable selection is mostly restricted to select the raw digitized voice signal as the variable; alternatively, the surrounding environmental noise could be used as another variable for noise cancelation purposes. Feature generation has been carried out by obtaining linear predictive coefficients (LPC),

AR/ARMA coefficients, Fourier coefficients, Cepstral coefficients, Mel Spectral Coefficients, and others [31]. In many the cases, the coefficients are computed over a short-time window (typically 10 ms) and over the voiced segments of the speech signal. As well as for the feature generation, for the classifier several options are available: Hidden Markov Models HHM (perhaps the most popular), neural networks (back propagation, self-organizing maps, radial basis, etc.), support vector and other kernel machines, Gaussian mixtures, Bayesian type classifiers, LVQ, and others. It should be noted that this list of choices of each element in the pattern recognition chain is by no means extensive. Please note that by simply considering all the combinations of variables, features and classifiers mentioned here, it is easily seen that there exist too many ways to implement a restricted-vocabulary speech recognition system.

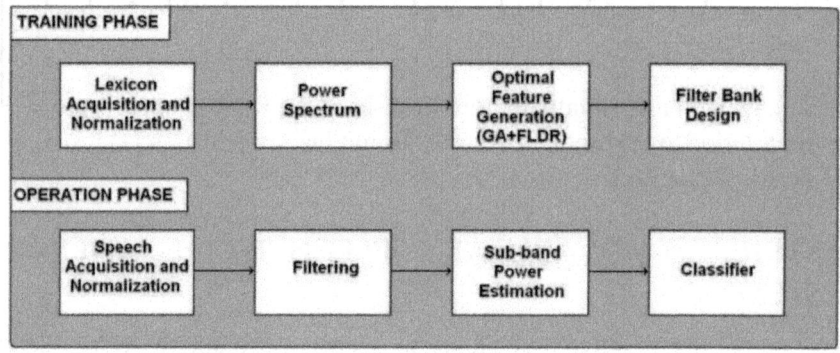

Figure. 8: Block diagram showing the optimal feature generation with the GA and FLDR.

METHODOLOGY

Figure 8 shows a block diagram of the whole speech recognition system. The system works in two phases: training and operation. The blocks of the training-phase are described below. Lexicon Acquisition and Normalization. In this stage, the set $L = \{w_0, w_1, w_2, ..., w_{M-1}\}$ of M words that will comprise the lexicon to be recognized is acquired from the speaker(s) that will use the system; vectors containing the digitized versions of the voice signals will be at the disposal of the next stage. Power Spectrum. The power spectrum has been traditionally the source of features for speech recognition; here, the power spectrum of the voice signals will be used by the genetic algorithm to determinate discriminant frequency bands. Optimal Feature Generation. The features selected here are a) the energy E of eight to twelve frequency regions (sub-bands) of the spectrum, b) the bandwidth (B_w) of each sub-band, and c) the central frequency (F_c) of the sub-bands. See Figureure 9. Sub-band processing in speech have been previously used, but in different manners [32], [33], [34]. A genetic algorithm with elitism is used here to select each bandwidth (BW) its central frequency (F_c) and a number of sub-bands. The main parameters of the genetic algorithm are listed in table 1. The cost function of the GA is an expression aimed at maximizing the Fisher's linear discriminant ratio (FLDR). The use of the FLDR in the cost function ensures

Table 1: Parameters of the Genetic Algorithm

Parameter	Value
Population	20-120
Mutation	Gaussian (1-2 , 1-2) (scale,shrink)
Selection	Elitism
Crossover	Scattered, one-point and two-points

increasing both, class separability and cluster compactness between classes $\omega 0$ and ω_1, being ω_0 the class of the word to be recognized and ω_1 the class of the rest of the M-1 words. The FLDR was described here in section 4.1, and the expression adopted here is the equation (31)

$$J_2 = \frac{|S_b|}{|S_w|} = |S_w^{-1} S_b|$$

where $|S_b|$ is the determinant of the inter-class covariance matrix and $|S_b|$ is the determinant of the intra-class covariance class. At the end of the evolutionary process the genetic algorithm produced a set of vectors with the parameters [E, BW, F_c] and the number of sub-bands, between 8 and 12.

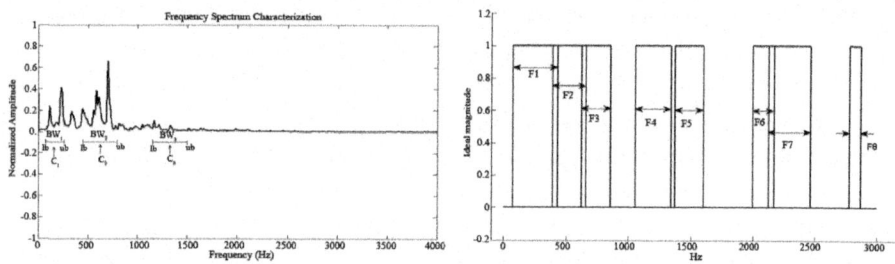

Figure. 9: Sub-band spectrum division to separate the power spectra in discriminant regions

Bank-filter Design. During the operation phase, the calculation of the power espectrum of the incoming voice signals is not practical for real-time response because the discrete Fourier transform (DFT [2], [3]) requires too much time to be calculated, and the fast Fourier Transform (FFT [2], [3]) requires that the number of samples to be a power of 2, requiring zero padding most of the times. Instead of using the power spectrum, in this application it is used the Parseval's theorem, described in section 4.2, which states that the energy in time and frequency is equal. In discrete form, we recall equation (32):

$$\sum_{i=0}^{N-1} x[i]^2 = \frac{2}{N} \sum_{k=0}^{N/2} |X[k]|^2$$

where x[i] is the time domain signal with and X[k] is its modified frequency espectrum, which is found by taking the DFT of the signal and dividing the first and last frequencies by the square rooth of two. Therefore, all that have to be done is to filter the sub-bands out of the time signals and then to calculate the energy in each sub-band. This can be perfectly accomplished by a bank of digital band-pass filters whose parameters match the parameters found by the genetic algorithm, and the advantage is that at the end of the last sample of voice in real-time a word can be immediately recognized.

In the Operation Phase, the three first stages operate simultaneously at each time a sample of the voice is acquired, and occur between two successive sampling intervals; in the first stage the voice signal is acquired and a normalization coefficient N_c is updated with the maximum value of the signal; the block labeled filtering represent the action of the filter bank and the outgoing signal is squared and added sample by sample. At the end of a voiced segment of sound, the third block provides to a classifier with the set of features for word classification.

System design

In this section the details of the system design will be presented.

Characterization of the frequency spectrum using sub-bands

Consider the Fourier Transform of a signal:

$$S(\omega) = F(s(t))$$

(40)

Now consider the frequency spectrum split in sub-bands, as shown in Figureure 9. Please notice in Figureure 9 that S(w) has been normalized to unitary amplitude. For each sub-band, the energy can be calculated as:

$$E_i = k_0 \sum_{lb}^{ub} \left(S(\omega)^2\right)$$

(41)

where E_i, is the energy of the i-th sub-band, k_0 is a constant proportional to the bandwidth (BW_i) of the i-th sub-band, l_b and u_b are the lower and upper bounds of the i-th sub-band. The feature vector of the n-th utterance of s(t) in the frequency domain becomes:

$$Sr_n(\omega) = [E_1 \ E_2 \ E_3 \ E_4 \cdots E_M]$$

(42)

In which M is the number of sub-bands, and S_m is the reduced version of the n-th S(ω). In order to characterize a word in the vocabulary, N samples of s(t) must be entered. From the example in Figureure 9 and without loss of generality, the set of parameters of the respective filter bank that will operate in the time domain is

$$BF = [C_1 \ BW_1 \ C_2 \ BW_2 \ C_3 \ BW_3 \cdots C_L \ BW_L]$$

(43)

The number of filters to be applied in the time-domain signal in this case is L. For a vocabulary of K words (K small), the spectra in which the solution must be searched is given by K matrices of N rows and 2000 columns.

GENETIC ALGORITHM SET UP

Coding the chromosome

The chromosome is comprised of the parameters of the filter bank described in equation 43. The number of sub-bands is fixed, and each one of the centers and bandwidths are subject to the genetic algorithm. Real numbers are used.

The cost function

The cost function is given by equation (32), criterion J2. The goal is to maximize J2 as a function of the centers and bandwidths.

Restrictions

The following restrictions apply:

R1. Sub-bands overlapping $<= 50Hz$,

R2. Bandwidth is limited to range from 40 to 400 Hz, varying according to the performance of the genetic algorithm.

Operating parameters of the genetic algorithm

The main parameters of the genetic algorithm are summarized in the Table 1. The values of the parameters are given according to the best results obtained by experimentation. The genetic algorithm was ran in Matlab®, using the genetic algorithms toolbox and the gatool guide. The nomenclature of the parameters in Table 1 is the one used by Matlab®.

Application's algorithm

To make operational de methodology described so far, the following steps apply:

1. Vocabulary definition. 2 to K words. In many real life applications, K in 8 to 15 words do the job.

2. Database acquisition. 15 to 20 utterances of each word from the vocabulary, for learning purposes. The sampling frequency can be set from 6000 to 8000 Hz. Human voice accommodates easily here.

3. s(t) to S(w) transformation. Apply Fourier transform to the data, normalize to unitary amplitude and to a fixed length of 2000.

4. Data preparation for the GA. Set-up the size (eq. 43) and restrictions (subsection 6.2.2.3) of the filter bank (chromosome).

5. Running the GA. Run the GA to find the sub-bands whose J_2 (eq. (32)) is the maximum.

6. Filters realization. For each sub-band, compute the coefficients of the respective bandpass filters. For real-time implementation, order from 4 to 8 is recommended, type IIR, elliptic. Elliptic filters achieve great

discrimination and selectivity. Implementation details can be found in [2].

7.   Modeling the commands. To make comparisons and therefore classification, a Gaussian statistical model of each word is to be constructed for each command in the vocabulary. Proceed as follows for each command:

   a.   Construct a matrix C of 15-20 rows of $S_m(w)$ and M columns (one row per sample of the command, one column per sub-band selected by the AG),

   b.   Compute the mean value overall the samples, to find the average energy per sub band, this is the feature vector of the command (μ_1)

   c.   Compute the covariance matrix, $S_i = cov(C_i)$.

   d.   At any time, the Mahalanobis distance between the model of the i-th command and the feature vector x of an incoming command is:

$$dM(x,\mu_i) = (x-\mu_i)^* \Sigma_i^{-1*}(x-\mu_i)^{-1}$$

(44)

To make the real-time implementation, a digital system must sample the input microphone continuously. Each sample of a command must be filtered by each filter (sub-band extraction). The integral of the energy of the signal leaving each filter over the period of the command is used to create the feature vector of that command, that is, x in equation 44. The feature vector is compared to each model in the vocabulary, and the command is recognized as the one with the minimum dM score. This is the so-called "minimum distance classifier" [5].

RESULTS

To test the system, the following lexicons were used:

   L_0={faster, slower, left, right, stop, forward, reverse, brake},

   L_1={zero, one, two, three, four, five, six, seven, eigth, nine},

   L_2={rápido, lento, izquierda, derecha, alto, adelante, reversa, freno},

   L_3 = {uno, dos, tres, cuatro, cinco}.

In all the lexicons, 3 male and 3 female volunteers were enrolled. They donated 116 samples of each word, 16 for training and 100 for testing. To demostrate the power of our approach we used the minimum distance classifier with the Mahalanobis distance. In all the cases, the genetic algorithm was ran 30 times to find the best response in the training set. During the training phase, the leave-one-out method was used to exploit the limited size of the training set [1]. Table 2 summarizes the results. In columns 5 and 7 are shown the comparison against a backpropagation neural network using as features Cepstral coefficients. The experiments were done using Matlab(R) and its associated toolboxes of genetic algorithms, neural networks and digital signal processing. The real-time implementation was done with a TMS320LF2407 Texas Instruments(R) Digital Signal Processor mounted on an experimentation card.

Table 2: Percentage (%) of correct classification with 4 lexicons, 2 languages, 6 persons, male and female voices, 2 classifiers. Simulations and real time implementation.

| | | | Simulations | | Real-time on DSP | |
| | | | MDC[3] | BPNN[4] | MDC[3] | BPNN[4] |
G[1]	L[2]	Training Set	Testing Set	Testing Set	Real scenario	Real scenario
Female	0	100	97	92	94	90
	1	100	98	90	95	90
	2	100	100	97	94	89
	3	100	100	91	95	88
Male	0	100	100	94	96	89
	1	100	100	90	95	90
	2	100	99	92	94	90
	3	100	98	92	93	88

[1]Gender, [2] Lexicon, [3]Minimum distance classifier, [4]Backpropagation neural network [1] 8-32-K neurons per layer, K according to the experiment, one neuron for each word in L.

Results and discussion

Results in the L3 Spanish vocabulary

The genetic algorithm was executed 30 times, and the maximum Fisher's ratio obtained was 62. The resulting best chromosome was:

BF = [254 180 526 132 744 118 1196 141 1483 115 2082 86 2295 171 2828 46]

From which the corresponding center and bandwidth were:

$C = [254\ 526\ 744\ 1196\ 1483\ 2082\ 2295\ 2828]$

$BW = [180\ 132\ 118\ 141\ 115\ 86\ 171\ 46]$

The recognition rates in the training and testing sets were 100% and 99%, respectively. In real conditions the correct classification rate was 93.5% in 40 repetitions of each word to the microphone.

The genetic algorithm in L_3

Consecutive executions of the GA produced variable Fisher's ratios. It was observed here that the population size is critical, since a population of 20 chromosomes produced Fisher's ratios between 35 and 42, while a population of 120 individuals easily produced Fisher's ratios above 58. It was noticed during experimentation that specific values for restrictions R1 and R2 also have a strong influence in the outcome. Comparing the results over a traditional approach with neural networks and cepstral coefficients it is evident a higher performance and, more important, the system exhibits realtime operation and very low computational effort compared to neural networks and realtime computation of the Cepstral coefficients.

Results in the L_0 English vocabulary

The genetic algorithm was executed 30 times, varying population size, probability of mutation, restrictions, number of sub-bands, and other parameters. The initial number of sub-bands was 8, then it was scaled to 9 and 12; in this scenario, the Fisher's ratio varied from 23 (8 sub-bands, population size = 20) to 48 (12 sub-bands, population size = 200). The resulting centers and bandwidths were:

$C = [250\ 446\ 648\ 1283\ 1483\ 1776\ 2018\ 2506\ 2737\ 3197\ 3383\ 3833]$

$BW = [5\ 138\ 187\ 157\ 139\ 43\ 207\ 106\ 105\ 98\ 148\ 224]$

Intra-class repeatability and inter-class differences

The performance in the training set was 100%; the performance in the testing set was 99%. The correct classification per word was {100% 100% 98% 100% 97% 100% 98% 99%} for the respective words {'faster','slower','left ','right','stop','forward','reverse','brake'}, respectively. Figureure 10 shows the normalized espectra of two utterances of the word 'faster' and the word 'slower'. In both cases notice the repeatability in the frequency domain, as well as the difference between both sets of spectra.

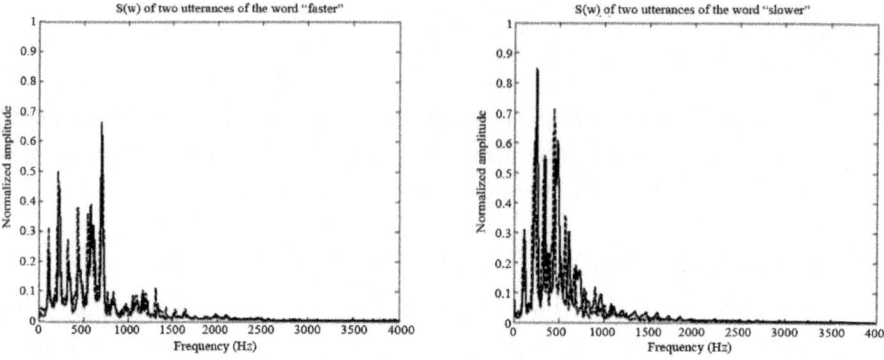

Figure. 10: Normalized spectra of the words "faster" and "slower".

Results of the real-time implementation

A minimum distance classifier was implemented in a digital signal processor TMS320LF2407 for each of the four lexicons L={L_0, L_1, L_2, L_3} from Texas Instruments, in order to verify the performance using in a nearly real-life application. The voiced/no-voiced segmentation was performed using a push-button to start and finish capturing the voice. The DSP has a built-in 10-bit analog to digital converter facilitating the interfacing task. The digital filters used were IIR topology, elliptic type, 8th order. The filter coefficients (A, B) were calculate using the Matlab® Software. The analog-to-digital conversion was set-up to acquire one sample every T seconds, (T=1/6000); each time a sample came into de device, the filters actuated and the respective output was squared and accumulated to calculate the energy of the signal. Scaling issues had to be solved since the model was created in a real valued [-1 , 1] scale, while the DSP just "see" integer values. Once a whole command was processed, it was just a matter of a few miliseconds to apply the minimum distance classifier and provide the classification. The correct classification rate was in this case of the order of 94.5%, in a total of 1200 repetitions of the words in L.

CONCLUSIONS AND FUTURE WORK

In this chapter was presented a method to implement a high performance, real-time, restricted-vocabulary speech recognition system, combining a genetic algorithm and the Fisher's Linear Discriminant Ratio (FLDR) in its matrix formulation. A review of the concepts of variable and feature selection as well as feature generation was made; also were presented some concepts related with speech processing, like the LPC formulation and the DTW method for template matching. One of the conceptual tools used here was the energy of the signal in certain sub-bands in the frequency domain; thanks to the Parseval's theorem, the same amounts of energy can be calculated in the time domain via a bank of digital filters, enabling thus a very fast way to apply the recognizer, since the process goes on at the same time as the occurrence of the word is exerted. Mainly, two experiments were shown, in Spanish and English, with male and female participants; in both cases high performance was attained, beyond 94% at the worst case. Compared to a typical implementation with backpropagation neural networks and cepstral coefficients, this approach was at least 10% more effective in near real-life application. The genetic algorithm consistently maximized the criteria of inter-class separability and intra-class compactness, under different conditions of population, probability of mutation, etc., and also varying the restriction set. It is remarkable and worth to mention that the genetic algorithm didn't gave the best result in its first execution, which means that the execution must be repeated to achieve good results; increasing the population and manipulating the restriction set demonstrated that it is possible to obtain a variety of different outcomes, so it is important to experiment carefully. The future work will consist of developing the voice/unvoiced detection in noisy environments, investigate an adapt more features that can be easily computed in time and with a dual in frequency, start working on the non-dependant speaker approach, making use of more robust classifiers, and finally, increase the vocabulary size, although still restricted to a specific semantic field, like in [35]. Another interesting venue is the one in which the user aging process is taken into account by the speech recognition system [36]

REFERENCES

1. K. Koutroumbas and S. Theodoridis, Pattern Recognition, 1st ed. California, E. U. A.: Academic Press, 1999.

2. John G. Proakis and Dimitris G. Manolakis, Digital Signal Processing. Principles, Algorithms, and Applications., 3rd ed. New Jersey, U.S.A.: Prentice Hall, 1996.

3. L. R. Rabiner and R. W. Schafer, Introduction to Digital Signal Processing, 1st ed. Hannover, U.S.A.: Now Publishers Inc., 2007.

4. Isabelle Guyon and André Elisseeff, An Introduction to Variable and Fature Selection, Journal of Machine Learning Research, vol. 3, pp. 1157-1182, 2003.

5. R. O. Duda, P. E. Hart, and D. G. Stork, Pattern Classification, 1st ed. New York, U.S.A.: John Wiley & Sons, Inc., 2001.

6. D. E. Goldberg, Genetic Algorithms in Search, Optimization and Machine Learning. U. S. A.: Addison-Wesley Professional, 1989.

7.   7 F. Itakura, Minimum Prediction Residual Principle applied to Speech Recognition, IEEE Transactions on Acoustics, Speech, and Signal Processing, vol. 23, no. 2, pp. 67-72, 1975.

8.   J. Makhoul, Linear Prediction: A Tutorial Review, Proceedings of IEEE, vol. 63, no. 4, pp. 561-580, 1975.

9.   C. D. Manning, Foundations of Statistical Natural Language Processing, 6th ed. Cambridge, Massachussets, U.S.A.: MIT Press, 2003.

10. 10 J. Ramírez, J. M. Górriz, and J. C. Segura, Voice Activity Detection. Fundamentals and Speech Recognition System Robustness, in Robust Speech Recognitin and Understanding, M. Grimm and K. Kroschel, Eds. Vienna, Austria: In-Tech, 2007, cap. 1, pp. 1-22.

11. L. D. Vignolo, H. L. Rufiner, D. H. Milone, and J. C. Goddard, Evolutionary Splines for Cepstral Filterbank Optimization in Phoneme Classsification, in EURASIP Journal on Advances in Signal Processing, vol. 2011, 2011, pp. 1-15.

12. T. Takiguchi, N. Miyake, H. Matsuda, and Y. Ariki, Voice and Noise Detection with AdaBoost, in Robust Speech Recognition and Understanding, M. Grimm and K. Kroschel, Eds. Vienna, Austria: In-Tech, 2007, cap. 4, pp. 67-74.

13. S. Y. Suk and H. Kojima, Voice Activated Appliances for Severely Disabled Persons, in Speech Recognition, Technologies and Applications, F. Mihelic and J. Zibert, Eds. Vienna, Austria: In-Tech, 2008, cap. 29, pp. 527-538.

14. R. Cardin, Improved Learning Strategies for Small Vocabulary Automatic Speech Recognition, McGill University, Montreal, Quebec, Canadá, Doctor of Philosophy Thesis 1993.

15. Isabelle Guyon. (2011, june) Isabelle Guyon's home page. on-line. Hyperlink http://www.clopinet.com/ isabelle/.

16. I. S. Oh, J. S. Lee, and B. R. Moon, Hybrid Genetic Algorithms for Feature Selection, IEEE Transactions on Pattern Analysis and Machine Intelligence, vol. 26, no. 11, pp. 1424-1437, November 2004.

17. P. R. Somol, J. Novovicova, and P. Pudil, Efficient Feature Subset Selection and Subset Size Optimization, in Pattern Recognition Recent Advances, A. Herout, Ed. Vienna, Austria: InTech, 2010, cap. 4, pp. 75-98.

18. Y. Sun, S. Todorovic, and S. Goodison, Local-Learning-Based Feature Selection for High-Dimensional Data Analysis, IEEE Transactions of Pattern Analysis and Machine Intelligence, vol. 32, no. 9, pp. 1610-1626, September 2010.

19. J. Teixeria de Souza, R. A. Ferreira do Carmo, and G. Campos de Lima, On the Combination of Feature Selection and Instance Selection, in Machine Learning, Y. Zhang, Ed. Vienna, Austria: In-Tech, 2010, cap. 9, pp. 158-171.

20. X. Huang, A. Acero, and H. W. Hon, Spoken Language Processing. A guide to Theory, Algorithm, and System Development, 1st ed. New Jersey, U.S.A.: Prentice Hall PTR, 2001.

21. A. Zacknich, Principles of Adaptive Filters and Self-Learning Systems, 1st ed. London, England: Springer, 2005.

22. 22 H. Sakoe and S. Chiba, Dynamic Programming Algorithm Optimization for Spoken Word Recognition , IEEE Transactions on Acoustics, Speech, and Signal Processing, vol. 26, no. 1, pp. 43-49, 1978.

23. P. J. Bigus and J. Bigus, Constructing Intelligent Agents with Java. A Programmer's Guide to Smarter Applications., 1st ed.: John Wiley & Sons, Inc., 2001.

24. K. S. Tang, K. F. Man, S. Kwong, and Q. He, Genetic Algorithms and their Applications, IEEE Signal Processing Magazine, pp. 22-37, November 1996.

25. S. M. Sait and A. Youssef, Iterative Computer Algorithms with Applications in Engineering., 1st ed. Los Alamitos, Cal., U.S.A.: IEEE Computer Society, 1999.

26. S. M. Ahadi, H. Sheikhzadeh, R. L. Brennan, and G. H. Freeman, An Effective FrontEnd for Automatic Speech Recognition, in 2003 International Conference on Electronics, Circuits and Systems, Sharah, United Arab Emirates, 2003, pp. 1-4, Subband speech recognition.

27. M. P. G. Saon, G. Zweig, J. Huang, B. Kingsbury, and L. Mangu, Evolution of the Performance of Automatic Speech Recogntion Algorithms in Transcribing

28. Conversational Telephone Speech, in IEEE Instrumentation and Measurement Technology Conference, Budapest, Hungary, 2001, pp. 1926-1931.

29. S. Kwong, Q. H. He, K. F. Man, K. S. Tang, and C. W. Chau, Parallel Genetic-based Hybrid Pattern Matching Algorithm for Isolated Word Recognition, International Journal of Pattern Recognition and Artificial Intelligence, vol. 12, no. 4, pp. 573-594, 1998.

30. V. V. Ngoc, J. Whittington, and J. Devlin, Real-time Hardware Feature Extraction with Embedded Signal Enhancement for Automatic Speech Recognition, in Speech Technologies. Vienna, Austria: In-tech, 2011, cap. 2, pp. 29-54.

31. S. A. Selouani and D. O'Shaughnessy, Robustness of Speech Recognition using Genetic Algorithms and Mel-cepstral Subspace Approach, in IEEE International Conference on Acoustics, Speech, and Singal Processing 2004 ICASSP '04, Montreal, Quebec, Canada, 2004, pp. I-201-4.

32. L. R. Rabiner, Tutorial on Hidden Markov Models and Selected Applications in Speech Recognition, Proceedings of the IEEE, vol. 77, no. 2, pp. 257-286, 1989.

33. S. Okawa, E. Bocchieri, and A. Potamianos, Multi-Band Speech Recognition in Noisy Environments, in International Conference on Acoustics, Speech and Signal Processing ICASSP 1998, Prague, Czech Republic, 1998, pp. 1-4.

34. A. de la Torre et al., Speech Recognition Under Noise Conditions: Compensation Methods, in Speech Recognition and Understanding, M. Grimm and K. Kroschel, Eds. Vienna, Austria: In-Tech, 2007, cap. 25, pp. 440-460.

35. A. Álvarez et al., Application of Feature Subset Selection Based on Evolutionary Algorithms for Automatic Emotion Recognition in Speech, Lecture Notes in Computer Science. Advances in Nonlinear Speech Processing, vol. 4885, pp. 273-281, May 2007.

36. G. E. Dahl, D. Yu, L. Deng, and A. Acero, Large Vocabulary Continuous Speech Recognition with Context Dependent DBN-HMMS, in International Conference on Acoustics, Speech and Signal Processing ICASSP 2011, Prague, Czech Republic, 2011, pp. 1-4.

37. B. M. Ben-David et al., Effects of Agging and Noise on Real-Time Spoken Word Recognition: Evidence from Eye Movements, Journal of Speech, Language, and Hearing Research, vol. 54, pp. 243-262, February 2011.

Chapter 7

PERFORMANCE OF VARYING GENETIC ALGORITHM TECHNIQUES IN ONLINE AUCTION

Kim Soon Gan, Patricia Anthony, Jason Teo and Kim On Chin

Universiti Malaysia Sabah, School of Engineering and Information Technology, Sabah Malaysia

INTRODUCTION

Genetic algorithm is one of the successful optimization algorithm used in computing to find ..exact or approximate solutions for certain complex problems. This novel algorithm was first ..introduced by John Holland in 1975 (Holland, 1975). Besides Holland, many other ..researchers have also contributed to genetic algorithm (Davis, 1987; Davis, 1991; ..Grefenstte, 1986; Goldberg, 1989; Michalewicz, 1992). This is an algorithm that imitates ..the evolutionary process concept based on the Darwinian Theory which emphasizes on ..the law of "the survival of the fittest". This algorithm used techniques which are inspired ..from evolution biology such as inheritance, selection, crossover and mutation ..(Engelbrecht, 2002). ..There are several important components in genetic algorithm which includes representation, ..fitness function, and selection operators (parent selection and survivor selection, crossover ..operator and mutation operator). Genetic algorithm starts by generating an initial ..population of individuals randomly. The individuals are represented as a set of parameter ..which is the solution to the problem domain. Normally, individuals are fixed length binary ..string. The individuals are then evaluated using fitness functions. The evaluation will give ..a fitness score to individuals indicating how well the solutions perform in the problem ..domain. The individuals that have been evaluated using the fitness function will be ..selected to be parents to produce offspring through the crossover and mutation operators. ..The genetic algorithms will repeat the above process except for the population ..initialization until the termination criteria is met.

Figure. 1 shows the structure of a genetic ..algorithm. ..GAs have been applied successfully in many applications including job shop scheduling ..(Uckun et al. 1993), the automated design of fuzzy logic controllers and systems (Karr 1991; ..Lee & Takagi, 1993), hardware-software co-design and VLSI design (Catania et al. 1997; ..Chandrasekharam et al. 1993). In this chapter, variations of genetic algorithms are applied in ..optimizing the bidding strategies for a dynamic online auctions environment. ..Auction is defined as a bidding mechanism and is expressed by a set of auction rules that ..specify how the winner is determined and how much he or she has to pay (Wolfstetter, ..2002). Jansen defines an online auction as an Internet-based version of a traditional auction ..(Jansen, 2003). In today's e-commerce market, online auction has acted as an important tool..in the services for procuring goods and items either for commercialize purposed or for ..personal used. Online auctions have been reported as one of the most popular and effective..ways of trading goods over the Internet (Bapna et al. 2001). Electronic devices, books, ..computer software, and hardware are among the thousands items sold in the online ..auctions every day. To date, there are 2557 auction houses that conduct online auctions as ..listed on the Internet (Internet Auction List, 2011). These auction houses conduct different ..types of auctions according to a variety of rules and protocols. eBay, as one of the largest ..auction house alone has more than 94 million registered users and had transacted more than ..USD 92 billion worth of goods during 2010 (eBay, 2010). These Figureures clearly show the .importance of online auctions as an essential for procuring goods in today's e-.commerce market method.

```
Begin
Generation = 0
Randomly Initialize Population
While termination criteria are not met
        Evaluate Population Fitness
        Crossover Process
        Mutation Process
        Select new population
        Generation = Generation + 1
End
```

Figure.1: The structure of a Genetic Algorithm.

The auction environment is highly dynamic in nature. Since there are a large number of..online auction sites that can be readily accessed, bidders are not constrained to participate in ..only one auction; they can bid across several alternative auctions for the same good ..simultaneously. As the number of auction increases, difficulties such as monitoring the ..process of auction, tracking bid and bidding in multiple auctions arise when the number ..of auctions increases. The user needs to monitor many auctions sites, pick the right ..auction to participate, and make the right bid in order to have the desired item. All of ..these tasks are somewhat complex and time consuming. The task gets even more ..complicated when there are different start and end times and when the auctions employ ..different protocols. For this reasons, a variety of software support tools are provided ..either by the online auction hosts or by third parties that can be used to assist consumers ..when bidding in online auctions. ..The software tools include automated bidding software, bid sniping software, and auction ..search engines. Automated bidding software or proxy bidders act on the bidder's behalf and ..place bids according to a strict set of rules and predefined parameters. Bid sniping software, ..on the other hand, is a practice of placing of bid a few minutes or seconds before an auction ..closes. These kinds of software, however, have some shortcomings. Firstly, they are only ..available for an auction with a particular protocol. Secondly, they can only remain in the ..same auction site and will not move to other auction sites. Lastly, they still need the ..intervention of the user, that is, the user still needs to make decision on the starting bid ..(initially) and the bid increments..To address the shortcomings mentioned above, an autonomous agent was developed that..can participate in multiple heterogeneous auctions. It is empowered with trading ..capabilities and it is able to make purchases autonomously (Anthony, 2003; Anthony & ..Jennings, 2003b). Two primary values that heavily influenced the bidding strategies of this ..agent are the k and β. These two values correspond to the polynomial function of the four ..bidding constraints, namely the remaining time left, the remaining auction left, the user's ..desire for bargain and the user's level of desperateness. Further details on the strategies will ..be discussed in Section 3. The k value ranges from 0 to 1 while the β value is from 0.005 to ..1000. The possible combinations between these two values are endless and thus, the search ..space for the solution strategies is very large. Hence, genetic algorithms were used to find ..the nearly optimal bidding strategy for a given auction environment. ..This work is an extension of the solution above, which has been successfully employed to ..evolve effective bidding strategies for particular classes of environment. This work is to ..improve the existing bidding strategy through the optimization techniques. Three ..different variations of genetic algorithm techniques are used to evolve the bidding ..strategies in order to search for the nearly optimal bidding solution. The three techniques ..are parameter tuning, deterministic dynamic adaptation, and self-adaptation. Each of this ..method will be detailed in Section 4, 5 and 6. The remainder of the chapter is organized as ..follow. Section 2 discusses related work. The bidding strategy framework is discussed in ..Section 3. The parameter tuning experiment is discussed in Section 4. Section 5 and 6 ..discussed the deterministic adaptive experiment and self-adaptive experiment. A ..comparison between all the schemes is discussed in Section 7. Finally, the conclusion is ..discussed in Section 8.

RELATED WORK

Genetic algorithm has shown to perform well in the complex system by which the old ..search algorithm has been solved. This is due to the nature of the algorithms that is able to ..discover optimal areas in a large search space with little priori information. Many researches ..in auctions have used genetic algorithm to design or enhance the auction's bidding ..strategies. The following section discusses works related to evolving bidding strategies. ...An evolutionary approach was proposed by Babanov (2003) to study the interaction of ..strategic agents with the electronic marketplace. This work describes the agents' strategies ..based on different methodologies that employ incompatible rules in collecting information ..and reproduction. This work used the information collected from the evolutionary ..framework for economic studies as many researches have attempted to use evolutionary ..frameworks for economics studies (Nelson, 1995; Epstein & Axtell, 1996; Roth, 2002; ..Tesfatsion, 2002). This evolutionary approach allows the strategies to be heterogeneous ..rather than homogenous since only a particular evolutionary approach is applied. This work ..has shown that the heterogeneous strategies evolved from this framework can be used as a ..useful research data. ..ZIP, introduced by Cliff, is an artificial trading agent that uses simple machine learning to ..adapt and operate as buyers or sellers in online open-outcry auction market environments ..(Cliff, 1997). The market environments are similar to those used in Smith's (Smith, 1962)...experimental economics studies of the CDA and other auction mechanisms. The aim of each..zip agent is to maximize the profit generated by trading in the market. A standard genetic ..algorithm is then applied to optimize the values of the eight parameters governing the ..behavior of the ZIP traders which previously must be set manually. The result showed that ..GA-optimized traders performed better than those populated by ZIP traders with manually ..set parameter values (Cliff, 1998a; Cliff, 1998b). This work is then extended to 60 parameters ..to be set correctly. The experiment showed promising result when compared to the ZIP ..traders with eight parameters (Cliff, 2006). Genetic algorithm is also used to optimize the ..auction market parameters setting. Many tests have been conducted on ZIP to improve the ..agent traders and the auction market mechanism using genetic algorithm (Cliff, 2002a; Cliff, ..2002b). Thus, ZIP was able to demonstrate that genetic algorithm can perform well in ..evolving the parameters of bidding agents and the strategies. ..In another investigation, a UDA (utility-based double auction) mechanism is presented ..(Choi et, al. 2008). In UDA, a flexible synchronous double auction is implemented where the ..auctioneer maximizes all traders' diverse and complex utility functions through ..optimization modeling based on genetic algorithm. It is a double auction mechanism based ..on dynamic utility function integrating the notion of utility function and genetic algorithm. ..The GA-optimizer is used to maximize total utility function, composed of all participants' ..dynamic utility functions, and matches the buyers and sellers. Based on the experimental ..result, it performance is better than a conventional double auction.

THE BIDDING STRATEGY FRAMEWORK

As mentioned, this work is an extension of Anthony's work (Anthony, 2003) to tackle the ..problem of bidding in multiple auctions that employ varying auctions protocols. This ..section details the electronic marketplace simulation, the bidding strategies and the genetic ..algorithm implemented in the previous work. .

The electronic market place simulation

.The market simulation employed three different auction protocols, English, Vickrey and ..Dutch that run simultaneously in order to simulate the real auction environment. The ..market simulation is used in this work to evaluate the performance of the evolved bidding ..strategies. The following section explains how the market simulation works. ..The marketplace simulator shown in Figure. 2 consists of concurrent running auctions that ..employ different protocols. These protocols are English, Dutch and Vickrey. All of these ..auctions have a known starting time and only English and Vickrey auctions have a known ..ending time. The bidding agent is given a deadline (t_{max}) by when it must obtain the desired ..item and it is told about the consumer's private valuation (p_r) for this item. The agent must .

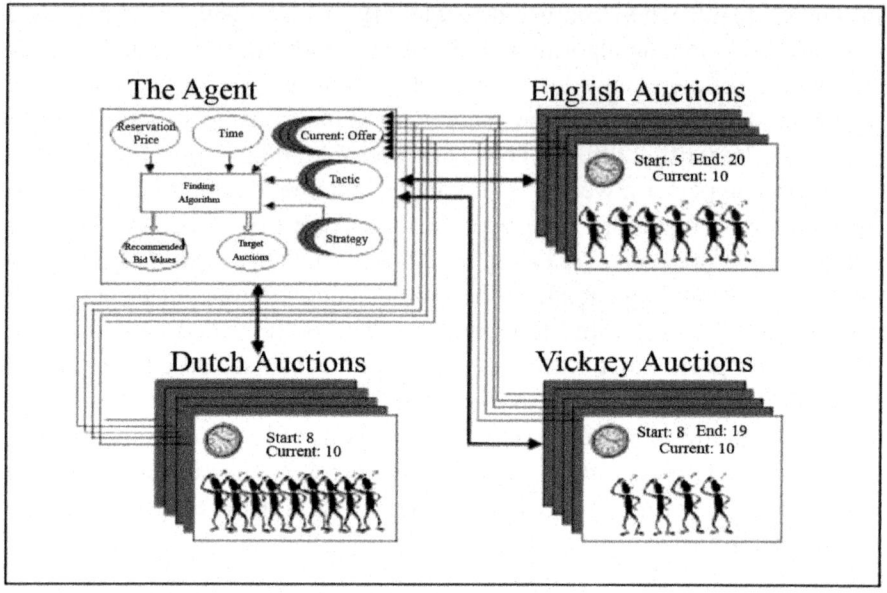

Figure. 2: The Marketplace Simulator

only buy an instance of the desired item.The marketplace announces the current bid values and the current highest bids for English ..auctions and the current offers for Dutch auctions at each time step. At the end of a given ..auction, it determines the winning bid and announces the winner. This set of information is .used by the agent when deciding in which auction to participate, at what value to bid and in ..which time to bid.

Bidding strategy

The bidding algorithm for this framework is shown in Figure. 3. Let Item_ NA be a boolean flag..to indicate whether the target item has already been purchased by the agent. Assume that ..the value of p_r is based on the current reliable market prices observed from past auctions ..and that the marketplace is offering the item which the agent is interested in. While the ..bidder agent has not obtained the desired item, the bidder agent needs to build an active ..auctions list in order to keep track of the current active auction. Active auction is defined as ..auction that is ongoing or just started but has not reach the ending time yet..

```
while (t ≤ tmax ) and (Item_NA = true)
    Build active auction list;
    Calculate current maximum bid using the agent's strategy;
    Select potential auctions to bid in, from active auction list;
    Select target auction as one that maximizes agent's expected utility;
    Bid in the target auction using current maximum bid as reservation price at this time;
Endwhile
```

Figure. 3:. The bidding agent's algorithm

```
for all i ∈ A
    if ((t ≥ σᵢ ) and (t ≤ ηᵢ ) or (Sᵢ (t) = ongoing)
        add i to L(t)
    endif
endfor
```

Figure. 4: Building active auction list algorithms

In order to build the active auction list, the bidder agent follows the algorithm as shown in Figure. 4. $S_i(t)$ is a boolean flag representing the status of auction $i \in A$, at time t, such that i ∈ A and $S_i(t)$ ∈ (ongoing; completed). Each auction i ∈ A, has a starting time σ_i, and its own ending time η_i. The active auction list is built by taking all the auctions that are currently running at time t. In English and Vickrey auctions, any auction that has started but has not reached its ending time is considered as active. $S_i(t)$ is used in Dutch auctions since the ending time of this type of auction is not fixed.

After the bidder agent builds the active auctions list, the bidder agent will start calculating the current maximum bid based on the agent strategy. The current maximum bid is defined as the amount of the agent willing to bid at the current time that is lesser than or equal to the agent's private valuation. Four bidding constraints are used to determine the current maximum bid namely the remaining time left, the remaining auction left, the desire for bargain and the level of desperateness. The remaining time tactic considers the amount of bidding time the bidder agent has to obtain the desire item. This tactic determines the bid value based on the bidding time left. Assuming that the bidding time t is between 0 and t_{max} $(0 \leq t \leq t_{max})$, the current bid value is calculated based on the following expression:

$$f_{rt} = \alpha_{rt}(t) P_r$$

(1)

where $\alpha_{rt}(t)$ is a polynomial function of the form:

$$\alpha_{rt}(t) = k_{rt} + (1 - k_{rt})\left(\frac{t}{t_{max}}\right)^{\frac{1}{\beta}}$$

(2)

This function is a time dependent polynomial function where the main consideration is the time left from the maximum time allocated. k_{rt} is a constant that determines the value of the starting bid of the agent in any auction multiplied by the size of the interval. This time dependent functions can be defined as those that start bidding near pr rapidly to those only bid near p_r right at the end along with all the possibilities in between with variation of the valueαrt (t). Different shapes of curve can be obtained by varying the values of β by using the equation defined above. There are unlimited numbers of possible tactics for each value of β. In this tactic, β value is defined between $0.005 \leq \beta \leq 1000$. It is possible to have two different behaviors for β. When β < 1, the tactic will bid with a low value until the deadline is almost reached, whereby this tactic concedes by suggesting the private valuation as the recommended bid value. When β > 1, the tactic starts with a bid value close to the private valuation and quickly reaches the private valuation long before the deadline is reached. Figure. 5 shows the different shape of the curves with varying β values.

The remaining auction left tactic, on the other hand, considers the number of remaining auctions that the bidder agent is able to participate in order to obtain the item. This tactic bids closer to p_r as the number of the remaining auctions decreases when the bidder agent is running out of opportunities to obtain the desired item. The current bid value is calculated based on the following expression:

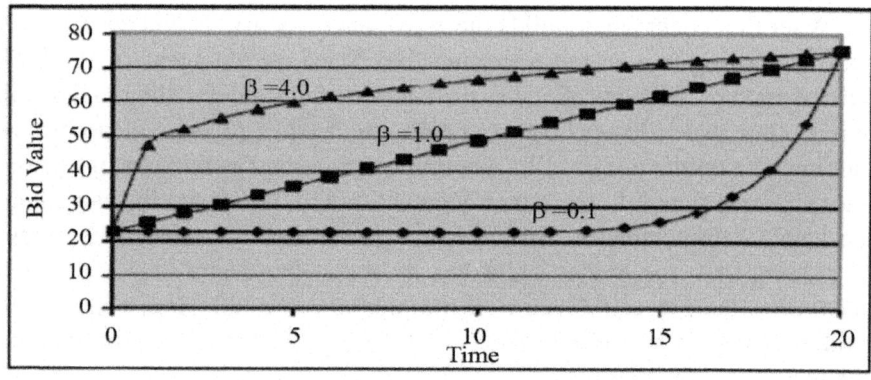

Figure. 5: The curve with varying β value. (Anthony, 2003)

$$f_{ra} = \alpha_{ra}(t)p_r$$

(3)

where $\alpha_{ra}(t)$ is a polynomial function of form:

$$\alpha_{ra} = k_{ra} + (1 - k_{ra})\left(\frac{c(t)}{|A|}\right)^{\frac{1}{\beta}}$$

(4)

The polynomial function αra is quite similar to the terms use in α_{rt}, whereby the only difference between the two function is the $c(t)$. $c(t)$ is a list of auctions that have been completed between time 0 and t. The β value for this tactic is identical to the remaining time tactic between $0.005 \leq \beta \leq 1000$.

The desire for a bargain tactic is the bidder agent that is interested in getting a bargain for obtaining the desired item. In this scenario, the bidder agent needs to take into account all the ongoing auctions and the time left to obtain the item. The current bid value is calculated based on the following expression:

$$f_{ba} = \omega(t) + \alpha_{ba}(t)(p_r - \omega(t))$$

(5)

In the expression above, the variable $\omega(t)$ takes into account all the ongoing auctions along with the current bid value. The Dutch and English are considered solely in this expression as only these two auctions have current bid value. As a consequence, the minimum bid value is calculated based on the current bid value and also the proportion of the time left in the auction. These values are summed and averaged with respect to the number of active auctions at that particular time. The expression for $\omega(t)$ is calculated based on the formula as below:

$$\omega(t) = \frac{1}{|L(t)|}\left(\sum_{1 \leq i \leq L(t)}\frac{t - \sigma_i}{\eta_i - \sigma_i}v_i(t)\right)$$

(6)

where $v_i i$ is the current highest bid value in an auction I at time t, and I at time t, and $I \in L(t)$; is the start and end time of auction i The expression for $\alpha_{ba}(t)$ is defined as:

$$\alpha_{ba}(t) = k_{ba} + (1 - k_{ba})\left(\frac{t}{t_{max}}\right)^{\frac{1}{\beta}}$$

(7)

The valid range for the constant k_{ba} is $0.1 \leq k_{ba} \leq 0.3$ and the β value is $0.005 \leq \beta \leq 0.5$. The β value is lower than 1 as bidder agent that is looking for bargain will never bid with the behavior of $\beta > 1$. The β value is, therefore, constantly lower than 1 in order to maintain a low bid until the closes to the end time. Hence, the value of $\beta < 0.5$ is used.

The level of desperateness tactic is the bidder agent's desperateness to obtain the target item within a given period and thus, the bidder agent who possesses this behavior tend to bid aggressively. This tactic utilizes the same minimum bid value and the polynomial function as the desire for bargain tactic but with a minor variation to the β and k_{de} value. The valid range for the constant k_{de} for this tactic is $0.7 \le k_{de} \le 0.9$ while the β value is $1.67 \le \beta \le 1000$. The β value is higher than 1 in this case as the bidder agent that is looking for bargain will never b_{id} with the behavior of β

There is a weight associated to each of this tactic and this weight is to emphasize which combination of tactics that will be used to bid in the online auction. The final current maximum bid is based on the combination of the four tactics by making use of the weight. Figure. 6 shows various combinations of the bidding constraints based on the different weight associated to the bidding tactics. It can also be seen that different bidding patterns are generated by varying the value of weights of the bidding constraints.

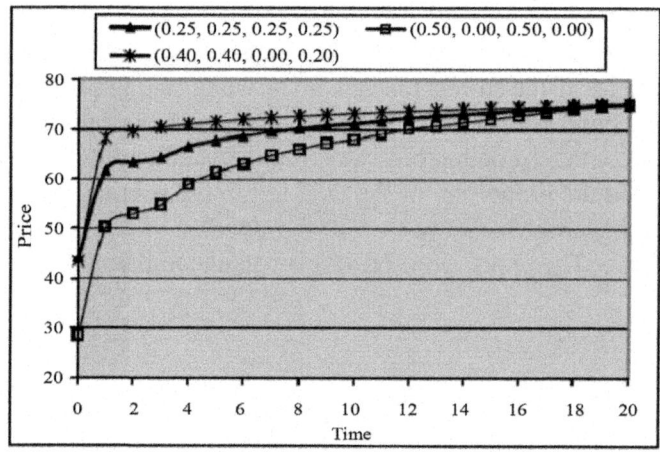

Figure. 6: Various combinations of the bidding constraints

GENETIC ALGORITHM

Representation

Floating point encoding is applied in this particular work as floating point encoding has shown to produce faster, more consistent and more accurate results (Janikow & Michalewicz, 1991). The floating encoding is, therefore, represented using an array of structure. The individuals that are represented in a floating point array structure are shown in Table 1.

Table 1. Bidding strategies representation

p_r	Agent's private valuation
t_{max}	Deadline given to the agent to obtain the desired item
k_{rt}	k for the remaining time tactic
β_{rt}	β for the remaining time tactic
k_{ra}	k for the remaining auction tactic
β_{ra}	β for the remaining auction tactic
k_{ba}	k for the desire for a bargain tactic
β_{ba}	β for the desire for a bargain tactic
k_{de}	k for the desperateness tactic
β_{de}	β for the desperateness tactic
w_{rt}	Relative weight for the remaining time tactic
w_{ra}	Relative weight for the remaining auction tactic
w_{ba}	Relative weight for the desire for a bargain tactic
w_{de}	Relative weight for the desperateness tactic
fitness	Fitness score for the individual

Representation

Fitness function is an objective function that quantifies the optimality of a solution in a genetic algorithm so that the particular chromosome may be ranked against all the other chromosomes. The main focus of the strategies evaluation in this work is the success rate and average utility of the strategies. Three fitness equations are used to evaluate the performance of the strategies namely the success rate, the agent's utility function and agent's utility with penalty. The success rate is the rate in obtaining the desired item and the second fitness function is the agent's utility

$$U_i(v) = \left(\frac{p_r - v}{p_r} \right) + c$$

(8)

where v represents the winning bid and c is an arbitrary constant 0.001 to ensure that the agent receives some value when the winning bid is equivalent to its private valuation. The third fitness equation involves a variation of the agent utility. If the agent fails to get the item, a penalty that ranges from 0.01 to 0.05 is incurred. Basically, Fitness Equation 1 is used if the delivery of the item is of utmost importance to the user. Fitness Equation 2 is used when the agent is looking for a bargain. Fitness Equation 3 is used when both the delivery of the item and looking for a bargain are equally important. The fitness score is then computed by taking the average utility from a total of 2000 runs.

Selection operators

Elitism is an operator used to retain some number of the best individuals in each generation to the next generation in order to ensure that the fittest individual is not lost during the evolution process (Obitko, 1998). Elitism is applied in this work to retain ten percent of the best individuals to the new population and to ensure that a significant number of the fitter individuals will make it to the next generation. Tournament selection is applied in the genetic algorithm for selecting the individuals to the mating pools for the remaining ninety percent of the population (Blickle & Thiele, 2001). Tournament selection technique was chosen because it is known to perform well in allowing a diverse range of fitter individuals to populate the mating pool (Blickle & Thiele, 1995). By implementing the tournament selection, fitter individuals can contribute to the next generation genetic construction and the best individual will not dominate in the reproduction process compared to the proportional selection

Crossover process

The extension operator floating point crossover operator is used this work (Beasley et al. 1993b). This operator works by taking the differences between the two values, adding it to the higher value (giving the maximum range), and subtracting it from the lower value (giving the minimum range). The new values for the genes are then generated between the minimum and the maximum range that were derived using this operator (Anthony & Jennings, 2002).

Mutation process

Since the encoding is a floating point, the mutation operator used in this work must be a non-binary mutation operator. Beasley has suggested a few non-binary mutation operators such as random replacement, creep operator and geometric creep (Beasley et al. 1993b) that can be used. The creep operator which adds or subtracts a small randomly generated amount from selected gene is used to allow a small constant of 0.05 to be added or subtracted from the selected gene depending on the range limitation of the parameter (Anthony & Jennings, 2002).

Stopping criteria

The genetic algorithm will repeat the process until the termination criteria are met. In this work, the evolution stops after 50 iterations. An extensive experiment was conducted to determine the point at which the population converges. It was decided to choose 50 as the stopping criterion since it is was observed that the population will always converge before or at the end of the 50 iterations. Anthony's work has some shortcoming where the crossover and mutation rate used in the work is based on literature review recommended values. However, researches have shown that the crossover rate and mutation rate applied in the application are application dependent, thus, simulation need to be conducted in order to find the suitable crossover and mutation rate. Besides that, other variations of genetic algorithm have proven to perform better that traditional genetic algorithm which is worthwhile to be investigated.

PARAMETER TUNING

Many researchers such De Jong, Grefenstte, Schaffer and others have contributed considerable efforts into finding the parameters values which are good for a number of

numerical test problems. The evolution of the bidding strategies by Anthony and Jennings (Anthony & Jennings, 2002) employed a fixed crossover and mutation probability based on the literatures. However, these recommended values may not perform at its best in the genetic algorithm as it has been proven that the parameter values are dependent on the nature of problems to be solved (Engelbrecht, 2002). In this experiment, the crossover and mutation rates are fine tuned with different combination of probabilities in order to discover the best combination of genetic operators' probabilities. Thus, the main objective of this experiment is to improve the effectiveness of the bidding strategies by "hand tuning" the values of the crossover rate and mutation rate to allow a new combination of static crossover and mutation rates to be discovered. By improving the algorithm, more effective bidding strategies can be found during the exploration of the solution. The experiment is subdivided to two parts. The first one varies the crossover rate and the second one varies the mutation rate. At the end of this experiment, the combination rate discovered is compared and empirically evaluated with the bidding strategies evolved in Anthony's work (Anthony, 2003).

EXPERIMENTAL SETUP

Table 2 and 3 show the evolutionary and parameter setting for the genetic algorithm. The parameters setting in the simulated environment for the empirical evaluations are shown in Table 4. These parameters include the agent's reservation price; the agent's bidding time and the number of active auctions. The agent's reservation price is the maximum amount that the agent is willing to pay for the item while the bidding time is the time allocated for the agent to obtain the user's required item. The active auctions are the list of auctions that is ongoing before time t_{max}. Figure. 7 shows the pseudocode of the genetic algorithm.

Table 2: Genetic algorithm evolutionary setting

Representation	Real Values Number
Crossover	Extension Combination Operator
Mutation	Creep Operator
Selection	Tournament Selection

Table 3: Genetic algorithm parameter setting

Number of Generations	50
Number of Individuals	50
Elitism	10%
Crossover Probability	0.2, 0.4, 0.6, 0.8
Mutation Probability	0.2, 0.02, 0.002
Termination Criteria	After 50 Generation
Number of Run	10

Table 4: ConFigureurable parameters for the simulated marketplace

Agent reservation price	$73 \leq p_r \leq 79$
Bidding time for each auction	$21 \leq t_{max} \leq 50$
Number of active auction	$20 \leq L(t) \leq 45$

```
Begin
    Randomly create initial bidder populations;
    While not (Stopping Criterion) do
        Calculate fitness of each individual by running the
        marketplace 2000 times;
        Create new population
        Select the fittest individuals (HP);
        Create mating pool for the remaining population;
        Perform crossover and mutation in the mating
        pool to create new generation(SF);
        New generation is HP + SF;
    Gen = Gen + 1
    End while
End
```

Figure. 7: Genetic algorithm

Experimental evaluation

The performance of the evolved strategies is evaluated based on three measurements. Firstly, the average fitness is the fitness of the population at each generation over 50 generations. The average fitness shows how well the strategy converges over time to find the best solution. Secondly, success rate is the percentage of time that an agent succeeds in acquiring the item by the given time at any price less than or equal to its private valuation. This measure will determine the efficiency of the agent in terms of guaranteeing the delivery of the requested item. Individual will be selected from each of the data set to compete in the simulated marketplace for 200 times. The success is calculated based on the number of time the agent is able to win the item over 200 runs. The formula below is used to calculate the success rate.

$$\text{Success Rate} = \frac{(\text{Number of winning}) \times 200}{100}$$

(9)

Finally, the third measurement is the average payoff which is defined as

$$\frac{\sum_{1 \leq x \leq 100} \left(\frac{p_r - v_i}{p_r} \right)}{n}$$

(10)

where p_r is the agent's private valuation, n is the number of runs, v_i is the winning bid value for auction i. This value is then divided by the agent's private valuation, summed and average over the number of runs. The agent's payoff is 0 if it is not successful in obtaining the item. A series of experiments was conducted using the set of crossover and mutation rate described in Table 2. It was found that 0.4 crossover rate and 0.02 mutation rate performed better than the other combinations (Gan et al, 2008a, Gan et al, 2008b). An experiment was conducted with the newly discovered crossover rate $p_c = 0.4$ and mutation rate $p_m = 0.02$. The result was then compared with the original combination of the genetic operators' ($p_c = 0.6$ and $p_m = 0.02$). Figureures 8, 9 and 10 shows the comparison between the strategies evolved using a combination of crossover rate 0.4 and a mutation rate of 0.02 and the combination of crossover rate 0.6 with a mutation rate of 0.02. The new strategies evolved from the combination of the crossover rate of 0.4 and mutation rate of 0.02 produced better result in terms of the average fitness, the success rate and the average payoff. It can be observed that the mutation rate of 0.02 evolved better strategies when compared to other mutation rates as well (0.2 and 0.002). This rate is similar to the research outcome by Cervantes (Cervantes & Stephen, 2006) in which a mutation rate below the 1/N and error threshold is recommended. Besides, the results of the comparison showed that the combination of 0.4 crossover rate and 0.02 mutation rate can achieve better balance in the exploration and exploitation in evolving the bidding strategies as well. T-test is performed to show the significant improvement of this newly discovered combination of genetic operator probabilities. The symbol of $\oplus$ in Table 5 indicates that the P-value is less than 0.05 and has significant improvement.

Table 5. P value of the t-test statistical analysis for comparison between newly discovered genetic operator probabilities with the old set of genetic operator probabilities

	P Value
Average Fitness	$\oplus$
Success Rate	$\oplus$
Average Payoff	$\oplus$

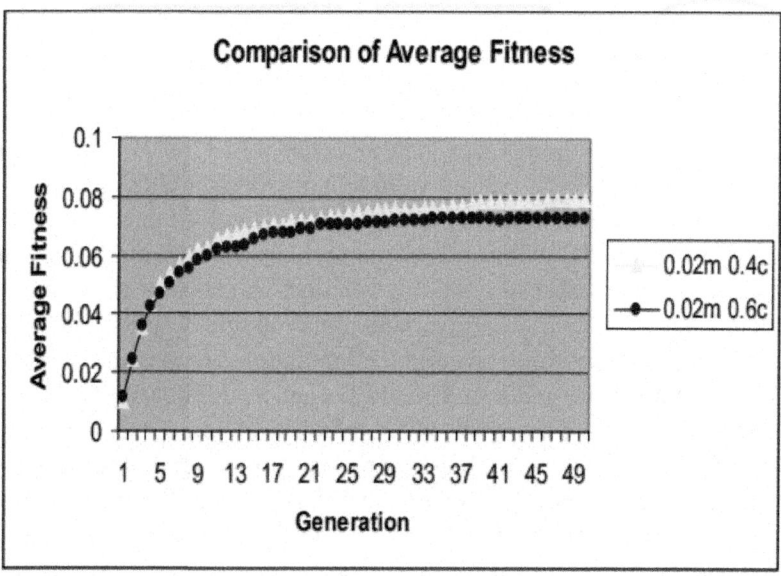

Figure. 8: Comparison of Average Fitness between the benchmark and the newly discovered rate.

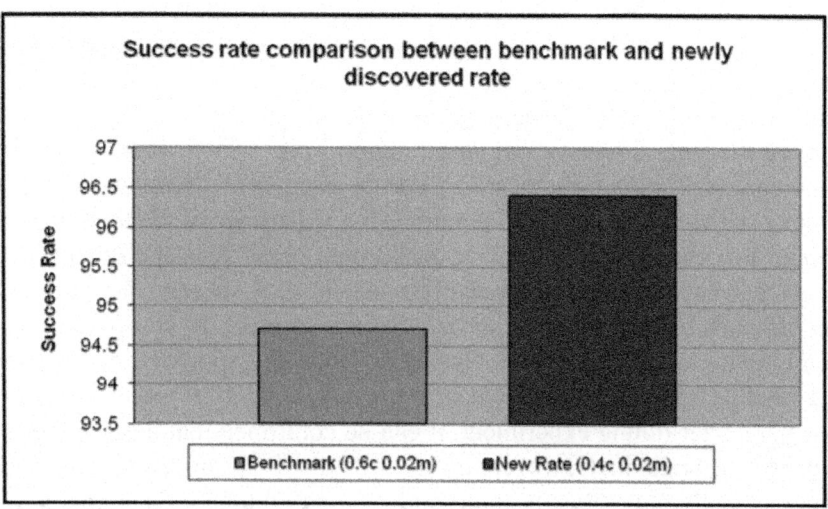

Figure. 9: Success rate for strategies evolved with the benchmark and the newly discovered rate

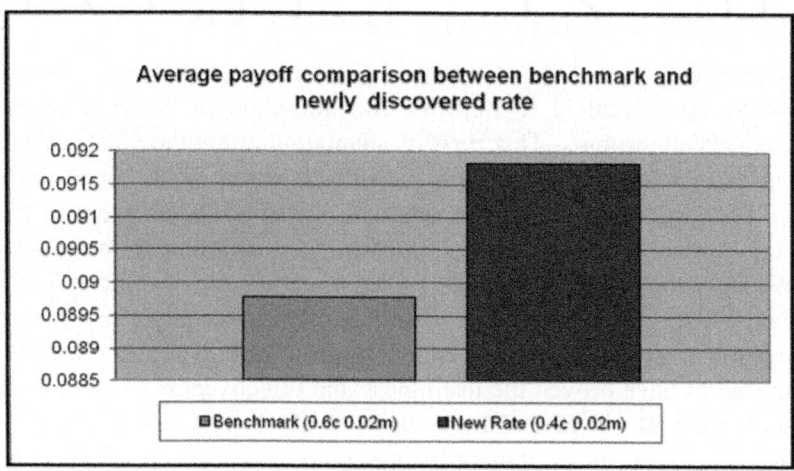

Figure. 10: Average payoff for strategies evolved with the benchmark and the newly discovered rate

This section investigated the performance of various combinations of predetermined sets of genetic operators' rates in genetic algorithm on a flexible and con Figureurable heuristic decision making framework that is capable to tackle the problem of bidding across multiple auctions that applied different protocols (English, Vickrey and Dutch). As mentioned earlier, the optimal combinations of operators' probabilities of applying these operators are problem dependent. Thus, experiments have to be conducted in order to discover a new operator of combinations genetic operator probability which can improve the effectiveness of the bidding strategy. This experiment has proven that the crossover rate and mutation rate which were applied in the previous work are not the best value to be used in this framework. With this new combination of genetic operators, the experimental evaluation has also shown that the strategies evolved performed better than the other strategies evolved from the other combinations in terms of success rate and average payoff when bidding in the online auction marketplace. By discovering a better combination of genetic operator's probabilities, the improved performance of the bidding strategies as shown in Figure. 8, 9, and 10 are achieved. From this parameter tuning experiment, it can be confirmed that the parameters are problem dependent. However, trying out all of the different combinations systematically is practically impossible as hand tuning the parameter is very time consuming. Therefore, in the second stage of the experiment, deterministic dynamic adaptation is applied to genetic algorithm to evolve the bidding strategies in order to overcome the manual tuning problem.

DETERMINISTIC DYNAMIC ADAPTATION

Many researchers have applied deterministic dynamic adaptation in evolutionary algorithms as a method to improve the limitation in the performance of evolutionary algorithms. This type of adaptation alters the value of strategy parameter by using some deterministic rule (Fogarty, 1989; Hinterding et al. 1997). The value of the strategy parameter is modified by the deterministic rule which is normally a time-varying schedule. It is different from the standard genetic algorithm since GA applies a fixed mutation rate over the evolutionary process. Most of the practical applications often favor larger or nonconstant settings of the genetic operators' probabilities. (Back & Schutz, 1996). Some of the studies have proved the usefulness and effectiveness of larger, varying mutation rates (Back, 1992; Muhlenbein, 1992). In this work, a time-variant dependent control rule is applied to change the control parameters over time without taking into account any present information by the evolutionary process itself (Eiben et al. 1999; Hinterding et al. 1997). Several studies have shown that a time dependent schedule is able to perform better than a fixed constant control parameter (Fogarty, 1989; Hesser & Manner, 1990; Hesser & Manner, 1992; Back & Schutz, 1996). The control rule is used to change the control parameter over the generation of the evolutionary process. The newly discovered crossover and mutation rates from the first experiment will be used in this particular schedule to serve as the midpoint in the time schedule. The parameter step size will change equally over the generation of the evolutionary process as well. This experiment is intended to discover the best deterministic dynamic adaptation by varying the genetic operators' probability scheme in exploring the bidding strategies. The deterministic increasing and decreasing schemes for the crossover and mutation are different due to the changing scale of the values. The newly discovered crossover rates obtained from Section 3 is used as the midpoint for the time variant schedule because the convergence period of the evolution occur around the 25th generation. Consequently, the deterministic increasing scheme for the crossover rate will change progressively from $p_c = 0.2$ to pc = 0.6 over the generation whereas the decrease scheme for the crossover rate is vice versa. The mutation rate obtained from the previous experiment is used as the midpoint of the time variant schedule for the increasing and decreasing schemes. The deterministic increasing scheme for the mutation rate, in contrast, will change progressively from $p_m = 0.002$ to pm = 0.2 over the generation and vice versa for the deterministic decreasing schemes. The changing scale during each generation is decided by taking the difference between ranges of the rate divided by the total number of generation.

Experimental setup

Table 6 shows the parameter setting for the deterministic dynamic adaptation genetic algorithm. The evolutionary setting and parameter setting in the simulated environment is the same as Tables 2 and 4. Figure. 11 shows the pseudocode of the deterministic dynamic adaptive genetic algorithm

Table 6: Deterministic dynamic adaptation parameter setting

Representation	Floating Points Number
Number of Generations	50
Number of Individuals	50
Elitism	10%
Selection Operator	Tournament Selection
Crossover Operator	Extension Combination Operator
Crossover Probability	Change(Range from 0.4 to 0.6) / Fixed (0.4)
Mutation Operator	Creep Operator
Mutation Probability	Change (Range from 0.2 to 0.002) / Fixed (0.02)
Termination Criteria	After 50 Generation
Numbers of Repeat Run	30

```
Begin
    Randomly create initial bidder populations;
    While not (Stopping Criterion) do
        Calculate fitness of each individual by running the
        marketplace 2000 times;
        Create new population
            Select the fittest individuals (HP);
            Create mating pool for the remaining population;
            Perform crossover and mutation in the mating
            pool to create new generation(SF);
            New generation is HP + SF;
        Change the control parameter value (Crossover / Mutation)
        Gen = Gen + 1
    End while
End
```

Figure. 11: The Deterministic Dynamic Adaptation Genetic Algorithm

Table 7: The Deterministic Dynamic Adaptation testing sets

Crossover Rate	Mutation Rate	Abbreviation
Fixed	Increase	CFMI
Fixed	Decrease	CFMD
Increase	Fixed	CIMF
Decrease	Fixed	CDMF
Increase	Increase	CIMI
Decrease	Decrease	CDMD
Increase	Decrease	CIMD
Decrease	Increase	CDMI

Experimental evaluation

The performance of the evolved bidding strategies is evaluated based on three measurements discussed in Section 4.2. As before, the average fitness of the each population is calculated over 50 generations. The success rate of the agent's strategy and the average payoff is observed over 200 runs in the market simulation. A series of experiments were conducted with the deterministic dynamic adaptation using the testing sets in Table 7. From the experiments, CFMD and CDMI performed better than the other combinations (Gan et al, 2008a, Gan et al, 2008b). Figure. 12 shows that the population evolved with deterministic dynamic adaptation is able to perform a lot better than the fixed constant crossover and mutation rates. This result is similar to the ones observed by other researches where non-constant control parameter performed better than fixed constant control parameter (Back 1992; Back 1993; Back & Schutz 1996; Fogarty 1989; Hesser & Manner, 1991; Hesser & Manner, 1992). Even though, the point of convergence for the different dynamic deterministic scheme is similar, the population with CDMI achieved a higher average fitness when compared to the populations with CFMD. The CDMI scheme with the increase mutation rate is able to maintain exploration velocity in the search space till the end of the run with the decreasing crossover rate achieving a balance between exploitation with the exploration in the search space and also to achieve a balance between exploration and exploitation in this setting.

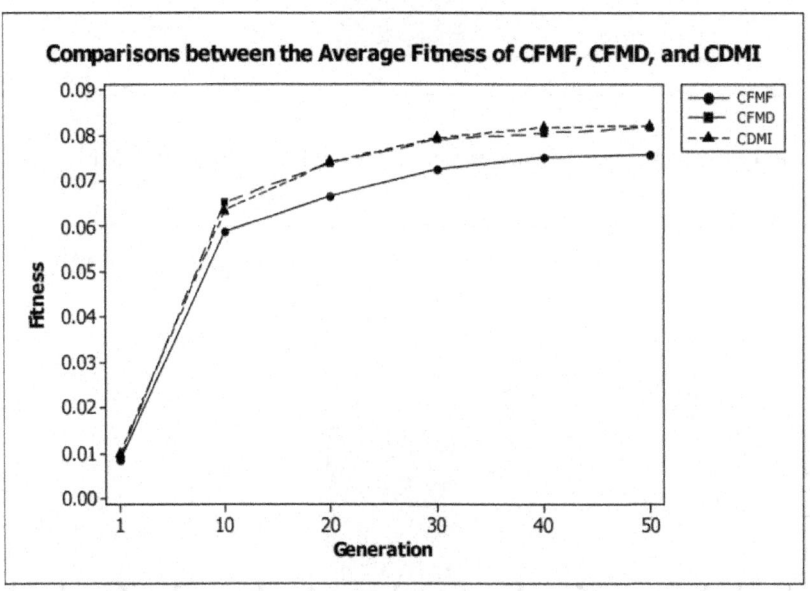

Figure. 12: Comparisons between the average fitness of CFMF, CFMD, and CDMI

Based on Figure. 13 and Figure. 14 CDMI outperformed CFMF and CFMD in both the success rate and the average payoff. This shows that the strategy evolved by using the CDMI does not only generate a better average fitness but also evolves better effective strategies compared to the strategy evolved for the other deterministic schemes and they are able to gain a higher profit when procuring the item at the end of the auction. It achieved a higher average fitness function during the evolution process as well.

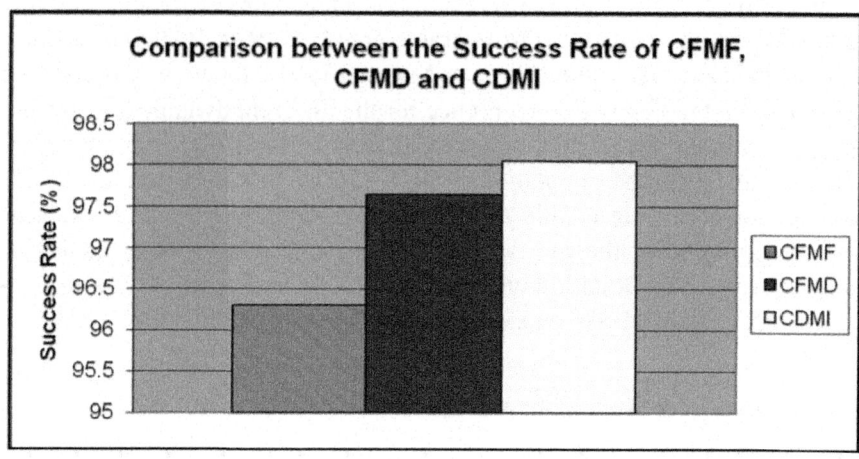

Figure. 13: Success rate comparison between CFMF, CFMD and CDMI

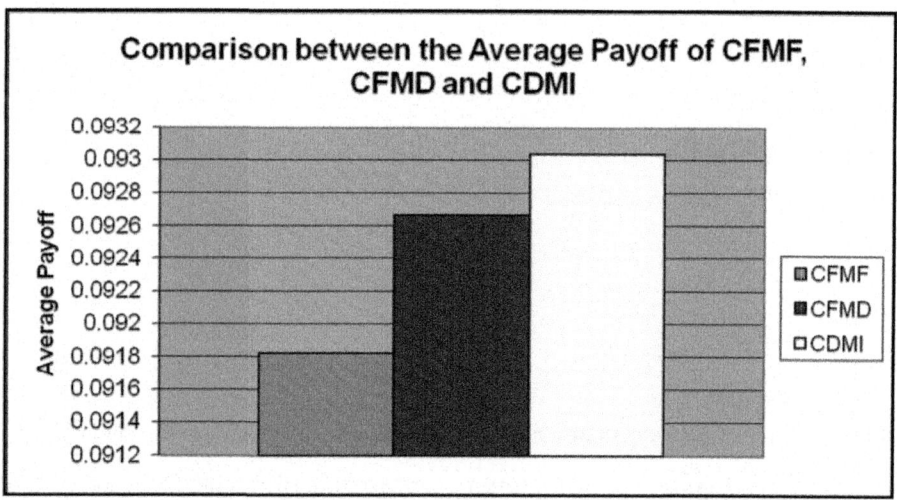

Figure. 14: Average payoff comparison between CFMF, CFMD and CDMI

This experiment has proven that non-constant genetic probabilities are more favorable than constant genetic probabilities. However, the deterministic dynamic adaptation may change the control parameter without taking into account the current evolutionary process as it does not take feedback from the current state evolutionary process whether the genetic operators' probabilities performed best at that current state of evolutionary process. The third stage of the experiment applies another adaptation method known as self adaptation. The self-adaptation method is different from the deterministic dynamic adaptation where the self-adaptation evolves the parameter based on the current status of the evolutionary process. The self-adaptation method incorporates the control parameters into the chromosomes, thereby, subjecting them to evolution. In the last stage of the experiment, the self-adaptation is applied to genetic algorithm in order to evolve the bidding strategies.

SELF-ADAPTATION

The idea of self-adaptation is based upon the evolving of evolution. Self-adaptation has been used as one of the method to regulate the control parameter. As the name implies, the algorithm controls the adjustment of the parameters itself. This is done by encoding the parameter into the individual genomes by undergoing mutation and recombination. The control parameters can be any of the strategy parameters in evolutionary algorithm such as mutation rate, crossover rate, population size, selection operators and others (Back et al. 1997). However, the encoded parameters do not affect the fitness of the individuals directly, but rather, "better" values will lead to "better" individuals and these individuals will be more likely to survive and produce offspring and hence, proliferating these "better" parameter values. The goal of the self-adaptation is not only to find the suitable adjustment but also to execute it efficiently. The task is further complicated when the optimizer faced by a dynamic problem is taken into account since a parameter setting that was optimal at the beginning of an EA-run might become unsuitable during the evolution process. This scenario has been shown in some of the researches that different values of parameters might be optimal at different stages of the evolutionary process (Back, 1992a; Back, 1992b; Back, 1993; Davis, 1987; Hesser & Manner, 1991). Self-adaptation aims at biasing the distribution towards appropriate regions of search space and maintains sufficient diversity among individuals in order to enable further evolvability (Angeline, 1995; Meyer-Nieberg & Beyer, 2006). The self-adaptation method has been commonly used in evolutionary programming (Fogel, 1962; Fogel, 1966) and evolutionary strategies (Rechenberg, 1973; Schwefel, 1977) but it is rarely used in genetic algorithms (Holland, 1975). This work applies self-adaptation in genetic algorithm which aims to adjust the crossover rate and mutation rate. The optimal rate for different phases of the evolution is obtained when different self-adaptation is capable in improving the algorithm by adjusting the crossover rate and mutation rate based on the current phase of the algorithm. Researchers have shown that the self-adaptation is able to improve the crossover in genetic algorithm (Schaffer & Morishima, 1987; Spears, 1995). In addition, studies also showed that the self-adaptive mutation rate does perform better than fixed constant mutation rate by incorporating the mutation rate into the individual genomes (Back, 1992a; Back, 1992b). In this section, three different self-adaptation schemes will be tested to discover the best self-adaptation scheme from this testing set. The self-adaptation requires the crossover and mutation rates to be encoded into the individual's genomes. Thus, some modification the encoding representation needs to be performed. The crossover and mutation rate become part of the genomes which will go through the crossover and mutation processes similar to the other alleles.

Experimental Setup

Table 8 shows the parameter setting for the self-adaptive genetic algorithm. The evolutionary setting and parameter setting in the simulated environment is same as Table 2 and 4. Figure. 15 shows the pseudocode of the deterministic dynamic adaptive genetic algorithm. Figure. 16 shows the different encoding representation of the individual genome that will be used in the experiment. The crossover and mutation rate are encoded into the representation in order to go through the evolution process.

Table 8: Self-adaptation genetic algorithm parameter setting

Representation	Floating Points Number
Number of Generations	50
Number of Individuals	50
Elitism	10%
Selection Operator	Tournament Selection
Crossover Operator	Extension Combination Operator
Crossover Probability	Self-Adapted / Fixed (0.4)
Mutation Operator	Creep Operator
Mutation Probability	Self-Adapted / Fixed (0.02)
Termination Criteria	After 50 Generation
Numbers of Repeat Run	30

```
Generation = 0
Random initialize population
While generation not equal 50
        Evaluate population fitness
        Select the top 10% to next generation
        Tournament Selection Parents to Mating Pool
        Check Parents Crossover Rate
        Generating offspring through crossover process
        Check Individual Mutation Rate
        Mutate the offspring
        Select offspring to the next generation
        Generation = Generation + 1
```

Figure. 15: The self adaptation algorithm both genetic operators

k_{rt}	β_{rt}	k_{ra}	β_{ra}	k_{ba}	β_{ba}	k_{de}	β_{de}	w_{rt}	w_{ra}	w_{ba}	w_{de}	p_c	p_m

Figure. 16: Encoding of a bidding strategy for self-adaptation crossover and mutation rate

Table 9: Self-adaptation testing sets

Crossover Rate	Mutation Rate	Abbreviation
Fixed	Self-Adapted	SAM
Self-Adapted	Fixed	SAF
Self-Adapted	Self-Adapted	SACM

Experimental Evaluation

The performance of the evolved bidding strategies is also evaluated based on the three measurements discussed in Section 4.2. As before, the average fitness of the each population is calculated over 50 generations. The success rate of the agent's strategy and the average payoff is observed over 200 runs in the market simulation. A series of experiments were conducted with the self-adaptive testing sets described in Table 10. From the experiments, self-adapting both crossover and mutation rates performed better than the other combinations (Gan et al, 2009). The population with self adaptive crossover and mutation

(SACM) achieved a higher average fitness compared to the population of self-adaptive crossover (SAC) and self –adaptive mutation schemes (SAM) as shown in Figure. 17. This scenario implies that the population with self adaptive crossover and mutation perform at its best among other populations and this is due to the self-adaptation crossover and mutation scheme which has combined the advantageous of the self-adaptive crossover and self-adaptive mutation scheme together. By having the two parameters to self-adapt, the control parameter can be adjusted to find the solution in different stages with the best control parameter which have been shown in the previous study indicating that different evolution stages will possess different optimal parameter values (Eiben et al. 1999).

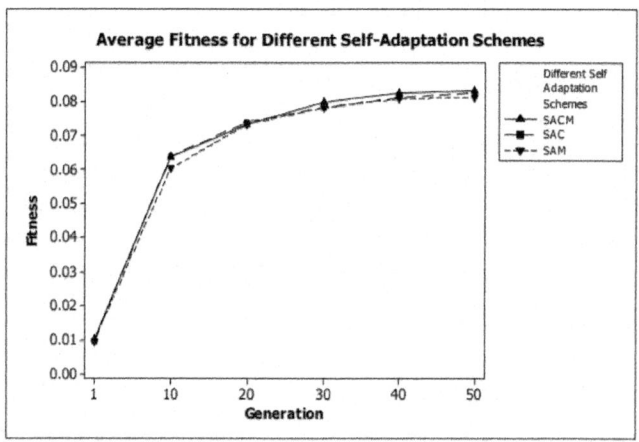

Figure. 17: Average fitness for different self-adaptation schemes

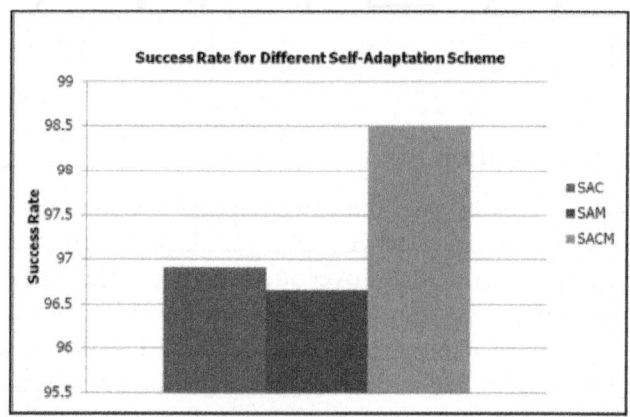

Figure. 18: Success rate for strategies evolved from different self-adaptation schemes

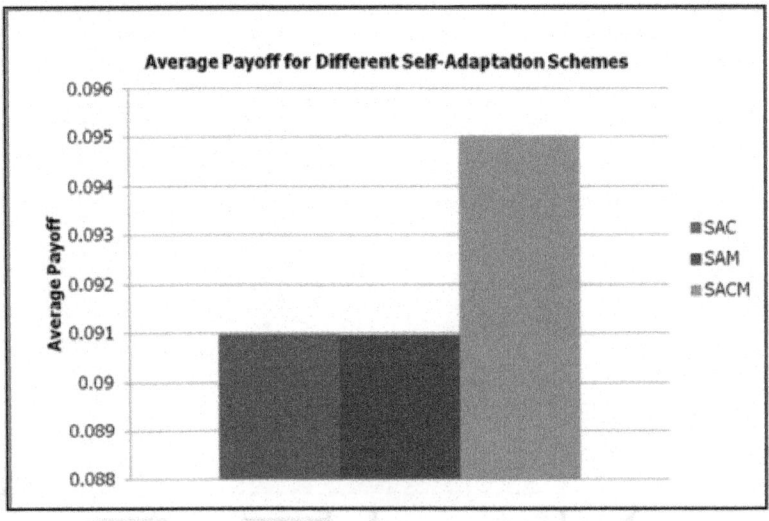

Figure. 19: Average payoff for strategies evolved from different self-adaptation schemes

All of the individuals generated a 4% increase in success rate and average payoff after employing the self adaptive crossover and mutation schemes as shown in Figure. 18 and Figure. 19 This has proven that the strategy evolved by using the self adaptive crossover and mutation does not only generate a better average fitness and success rate but also evolves better effective strategies compared to the strategy evolved for other self adaptive schemes.

COMPARISON BETWEEN VARIATIONS OF GENETIC ALGORITHM

In order to determine which of the three approaches perform the best in improving the effectiveness of the bidding strategies, the best result of each experiment is compared. The comparison is made by choosing the best performing schemes from the parameter tuning, deterministic dynamic adaptation and self-adaptation experiments. The main objective of this work is to improve the effectiveness of the existing bidding strategies by using different disciplines of the genetic algorithm.

Figure. 20 shows the average fitness for the evolving bidding strategy with different disciplines of the genetic algorithm. It can be seen clearly that there is an obvious differences between the convergence points in the different genetic algorithm disciplines. Self-adaptation

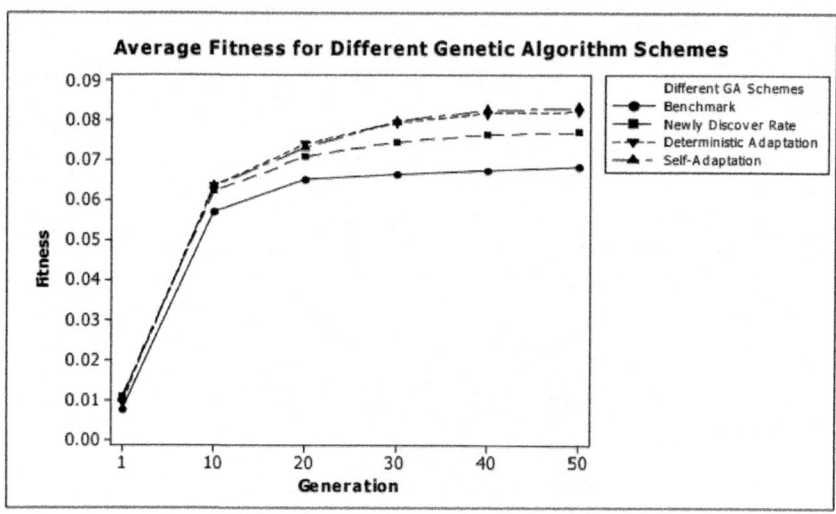

Figure. 20: Average fitness population with different genetic algorithm disciplines

achieves a higher average fitness compared to benchmark, the newly discovered static rate and deterministic dynamic adaptation. Although average fitness of the self-adaptation and deterministic dynamic adaption is similar, self-adaptation achieves a higher average fitness when compared to deterministic dynamic adaptation.

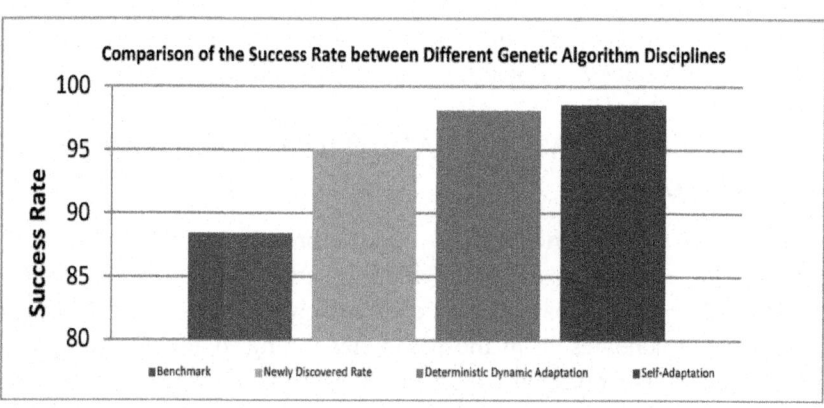

Figure. 21: Success rate for strategies evolved from different genetic algorithm disciplines

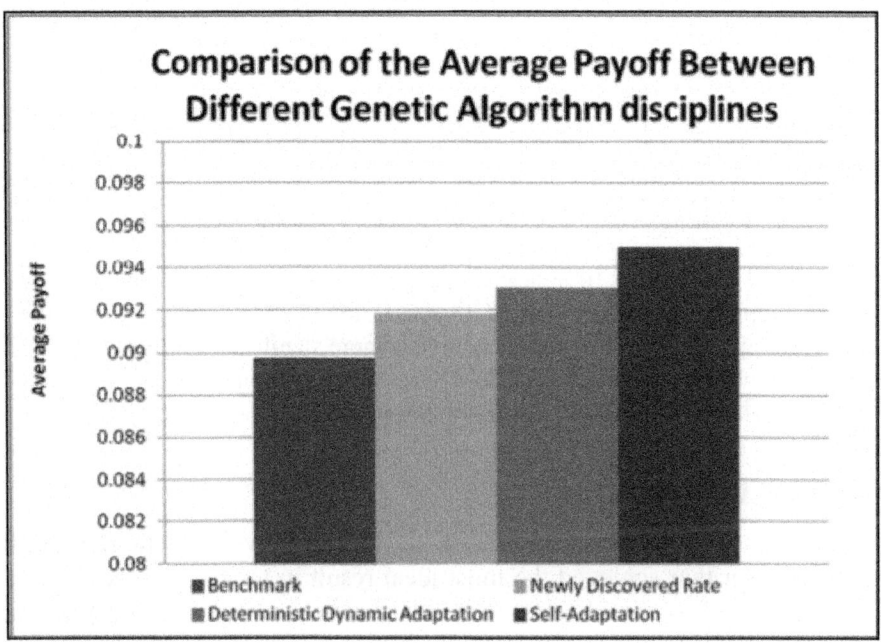

Figure. 22: Average payoff for strategies evolved from different genetic algorithm disciplines

The individuals evolved from the self adaptive genetic algorithm outperformed the other individuals from the other disciplines by delivering a more promising success rate. The strategy evolved is 1% higher than the strategies evolved from the deterministic dynamic adaptation. When compared to the benchmark value, an increase of 4% in the success rate is generated by the strategy which that employed the self-adaptation method. As a result, the strategy evolved from the self adaptive genetic algorithm can evolve better strategies and deliver higher success rate when bidding in online auctions which will eventually, improve the GA in searching for better bidding strategies. All of the strategies evolved from the self adaptive genetic algorithm outperformed the rest with 2% higher average payoff when compared to the strategies which applied deterministic dynamic adaptation and 4% higher when compared to the strategies from the benchmark. This result obtained indicates that the strategy evolved by using the self adaptive genetic algorithm does not only produce a better average fitness and success rate but also evolves better effective strategies compared to the other strategies evolved for other disciplines and they have gained higher profit when procuring the item.

Table 10. P value for the comparison between different disciplines in term of success rate and average payoff

SA	Benchmark	Newly Discovered Static Rate	DDA
Success Rate	⊕	⊕	⊕
Average Payoff	⊕	⊕	⊕

The symbol ⊕ in Table 10 indicates that the P-value is less than 0.05 and has significant improvement. The result of P value in the t-test in Table 10 shows the improvement generated by the self-adaptation is more significant compared to the other disciplines. Hence, it can be confirmed that self-adaptation is the best discipline in improving the effectiveness of the bidding strategies.

CONCLUSION

Based on the results of the experiments, the strategies evolved with self adaptive genetic algorithm achieved the most ideal result in terms of success rate and average payoff in an online auction environment setting. The strategies have also achieved a higher average fitness function during the evolution process. The result in Figureure 20, 21, 22 and Table 10 confirmed this conclusion by empirically proving that self adaptive genetic algorithm can evolve better bidding strategies compared to the other genetic algorithm disciplines. Among these different methods, the self-adaptation outperformed all of the other methods due to the nature of the method. In order to achieve better bidding strategies, the self-adaptation crossover and mutation scheme can be used to ensure better bidding strategies which in turn produces higher success rate, average fitness and average payoff. Further investigation can be conducted by evolving the bidding strategies with two other evolution methods which are the evolution strategies and evolution programming. Evolving the bidding strategies with the evolution programming and evolution strategies may generate interesting result which different from genetic algorithm. A comparison between performances the evolutions strategies, evolution programming and genetic algorithm may produce interesting results.

REFERENCES

1.   Angeline, P. J. 1995. Adaptive and Self-Adaptive Evolutionary Computation. In Palaniswami, M., Attikiouzel, Y., Marks, R. J., Fogel, D., & Fukuda, T. (eds.). A Dynamic System Perspective, pp. 264-270. New York: IEEE Press.

2.   Anthony, P. 2003. Bidding Agents for Multiple Heterogeneous Online Auctions. PhD's Thesis. University of Southampton.

3.   Anthony, P. and Jennings, N. R. 2002. Evolving Bidding Strategies for Multiple Auctions. Amsterdam: Proceedings of the 15th European Conference on Artificial Intelligence, pp. 178-182. IOS Press.

4.   Anthony, P. and N. R. Jennings. 2003a. Agents in Online Auctions. In Yaacob, S. Nagarajan, R., Chekima, A. and Sainarayanan, G. (Eds.). Current Trends in Artificial Intelligence and Applications, pp. 42-50.

5.   Kota Kinabalu: Universiti Malaysia Sabah. Anthony, P. and N. R. Jennings. 2003b. Developing a Bidding Agent for Multiple Heterogeneous Auctions. ACM Transactions on Internet Technology, 3(3): 185-217. Anthony, P. and N. R. Jennings. 2003c.

6.   A Heuristic Bidding Strategy for Multiple Heteregenous Auctions. Proceedings of the Fifth International Conference on Electronic Commerce, pp. 9-16. New York:

7.   ACM. Babanov, A., Ketter, W. and Gini, M. L. 2003. An Evolutionary Approach for Studying Heterogeneous Strategies in Electronic Markets.

8.   Engineering Self-Organising Systems 2003. pp. 157-168. Back T. and Schutz M. 1996. Intelligent Mutation Rate Control in Canonical Genetic

9.   Algorithms. Proceedings of the International Symposium on Methodologies for Intelligent Systems. In Ras, Z. W. and Michalewicz, Z. (Eds.) Lecture Notes In Computer Science, 1079: 158-167.

10. Back, T. 1992a. The Interaction of Mutation Rate, Selection, and Self-Adaptation within a Genetic Algorithm. In Manner, R. and Manderick, B. (Eds). Proceeding 2nd Conferences of Parallel Problem Solving from Nature, pp. 85-94. Belgium: Elsevier.

11. Back, T. 1992b. Self-Adaptation in Genetic Algorithms. In Varela, F. J. and Bourgine, P. (Eds.) Toward a Practice of Autonomous Systems: Proceeding 1st European Conference of Artificial Life, pp. 263-271. Cambridge: MIT Press.

12. Back, T. 1993. Optimal Mutation Rates in Genetic Search. Proceedings of the 5th International Conferences of Genetic Algorithms, pp. 2-8. San Francisco: Morgan Kaufmann.

13. Back, T. Fogel, David. and Michalewicz, Z. Eds. 1997. Handbook of Evolutionary Computation. New York: Oxford University Press.

14. Bapna, R., P. Goes, and A. Gupta (2001). Insights and Analyses of Online Auctions. Communications of the ACM, 44 (11): 43-50.

15. Beasley, D., Bull, D. R. and Martin R. R. 1993. An Overview of Genetic Algorithms: Part 2, Research Topics. University Computing 15(4): 170 - 181.

16. Blickle, T. and Thiele, L. 1995. A Comparison of Selection Schemes Used in Genetic Algorithms. Technical Report 11. Zurich: Swiss Federal Institute of Technology.

17. Blickle, T. and Thiele, L. 2001. A Mathematical Analysis of Tournament Selection. Proceedings of the Sixth International Conference on Genetic Algorithms, pp. 9-16. San Francisco: Morgan Kaufmann.

18. Catania, V., Malgeri, M. and Russo, M. 1997. Applying Fuzzy Logic to Codesign Partitioning. IEEE Micro 17(3): 62-70. Cervantes, J. and Stephens, C. R. 2006. "Optimal" mutation rates for genetic search. Proceedings of the 8th Annual

19. Conference on Genetic and Evolutionary Computation Conference, pp.1313 – 1320. New York: ACM Press.

20. Chandrasekharam, R., Subhramanian, S. and Chaudhury, S. 1993. Genetic Algorithm for Node Partitioning Problem and Application in VLSI Design. IEE Proceedings Series E: Computers and Digital Techniques, 140(5): 255–260.

21. Choi, J. H., Ahn, H., and Han, I. 2008. Utility-based double auction mechanism using genetic algorithms. Expert System Applicationl. 2008, pp. 150-158.

22. Cliff, D. 1997. Minimal Intelligence Agents for Bargaining Behaviours in Market Environment. Technical Report HPL-97-91. Hewlett Packard Laboratories.

23. Cliff, D. 1998a. Genetic optimization of adaptive trading agents for double-auction markets. Proceedings Computing Intelligent Financial Engineering (CIFEr). pp. 252–258.

24. Cliff, D. 1998b. Evolutionary optimization of parameter sets for adaptive software-agent traders in continuous double-auction markets. Artificial Society Computing Markets (ASCMA98) Workshop at the 2nd Int. Conference. Autonomous Agents. (unpublished)

25. Cliff, D. 2002a. Evolution of market mechanism through a continuous space of auction types. Proceeding Congress Evolutionary Computation. pp. 2029–2034

26. Cliff, D. 2002b. Visualizing search-spaces for evolved hybrid auction mechanisms. Presented at the 8th Int. Conference. Simulation and Synthesis of Living Systems (ALifeVIII) Conference. Beyond Fitness: Visualizing Evolution Workshop, Sydney.

27. Cliff, D. 2006. ZIP60: Further Explorations in the Evolutionary Design of Trader Agents and Online Auction-Market Mechanisms. IEEE Transactions on Evolutionary Computation.

28. Davis, L. 1987. Genetic Algorithm and Simulated Annealling. San Francisco: Morgan Kaufmann. Davis, L. 1991a. Handbook of Genetic Algorithms. New York: Van Nostrand Reinhold.

29. Davis, L. 1991b. Hybridization and Numerical Representation, In

30. Davis, L. (ed), The handbook of Genetic Algorithm, pp. 61-71. New York: Van Nostrand Reinhold. eBay. 2008. "eBay Inc. Annual Report 2010," (19 October 2010). http://investor.ebayinc.com/annuals.cfm ..

31. Eiben, A. G., Hinterding, R., and Michalewicz, Z. 1999. Parameter Control in Evolutionary Algorithms. IEEE Transactions on Evolutionary Computation, 3(2). pp. 124 – 141.

32. Engelbrecht, A.P. 2002. Computational Intelligence an Introduction. New Jersey: John Wiley & Sons.

33. Epstein, J. M. and Axtell, R. 1996. Growing Artificial Societies: Social Science from the Bottom Up. Cambridge: MIT Press.

34. Fogarty, T. 1989. Varying the probability of mutation in genetic algorithm. In Schaffer, J. D. (Ed.) Proceedings of the Third International Conference on Genetic Algorithms, pp. 104- 109. San Francisco: Morgan Kaufmann.

35. Fogel, D. B. 1992. Evolving Artificial Intelligence. PhD Thesis. Berkeley: University of California.

36. Fogel, L. J. 1962. Autonomous Automata. Industrial Research, 4: 14-19. Gan K.S., Anthony P. and Teo J. 2008a. The Effect of Varying the Crossover Rate in the Evolution of Bidding Strategies. 4th International IASTED Conference on Advances in Computer Science and Technology (ACST-2008), Langkawi, Malaysia, April 2008.

37. Gan K.S., Anthony P., Teo J. and Chin K.O. 2008b, Mutation Rate in The Evolution of Bidding Strategies, The 3rd International Symposium on Information Technology 2008 (ITSim2008), Kuala Lumpur, Malaysia, August 2008

38. Gan K.S., Anthony P., Teo J. and Chin K.O. 2008c, Dynamic strategic parameter control in evolving bidding strategies. Curtin University of Technology Science and Engineering (CUTSE) International Conference 2008, Sarawak, Malaysia, November 2008.

39. Gan K.S., Anthony P., Teo J. and Chin K.O. 2008d, Evolving Bidding Strategies Using Deterministic dynamic adaptation. The 4th International Conferences on Information Technology and Multimedia (ICIMU2008), Bangi, Malaysia, November 2008

40. Gan K.S., Anthony P., Teo J. and Chin K.O. 2009, Evolving Bidding Strategies Using SelfAdaptation Genetic Algorithm, International Symposium on Intelligent Ubiquitous Computing and Education, Chengdu, China.

41. Goldberg, D. E. 1989. Genetic Algorithms in Search, Optimization and Machine Learning. New York: Addison-Wesley.

42. Hesser, J. and Manner, R. 1990. Towards an optimal mutation probability for genetic algorithms. In Schewefel, H. P. and Manner, R. (Eds.) Proceedings for eh 1st Conferences on Parallel Problem Solving from Nature, Lecture Notes in Computer Science, 496, pp 23-32. London: Springer-Verlag.

43. Hesser, J. and Manner, R. 1992. Investigation of the m-heuristic for optimal mutation probabilities, Proceedinng of the 2nd Parallel Problem Solving from Nature. pp. 115-124. Belgium: Elsevier.

44. Hinterding, R., Michalewicz, Z., and Eiben, A. E. 1997. Adaptation in Evolutionary Computation: A survey. Proceeding 4th IEEE Conference of Evolutionary Computation. pp. 65-69.

45. Hinterding, R., Michalewicz, Z., and Eiben, A. E. 1997. Adaptation in Evolutionary Computation: A survey. Proceeding 4th IEEE Conference of Evolutionary Computation. pp. 65-69.

46. Holland, J. H. 1975. Adaption in Natural and Artificial System. Michigan: MIT Press. Internet Auction List. 2008. Listing Search in USAWeb.com, http://internetauctionlist.com/Search.asp. 19 October 2011.

47. Janikow, C. Z. and Michalewiz, Z. 1991. An experimental comparison of Binary and Floating Point Representations in Genetic Algorithms, In Belew, R. K. and Booker, L. B. (eds), Proceedings of the 4th International Conferences in Genetic Algorithms. pp 31-36. San Francisco: Morgan Kaufmann.

48. Jansen, E. 2003. Netlingo the Internet Dictionary. http://www.netlingo.com/. 10 November 2008.

49. Karr, C. 1991. Genetic Algorithms for Fuzzy Controllers. AI Expert. 6(2): 26-33. Lee, M. A. and Takagi, H. 1993. Integrating Design Stages of Fuzzy Systems Using Genetic Algorithms. Proceedings of the IEEE International Conference on Fuzzy Systems. pp. 612–617.

50. Meyer-Nieberg, S. and Beyer, H-G. 2006. Self-Adaptation in Evolutionary Algorithms. In Lobo, F., Lima, C., and Michalewicz, Z. (Eds.) Parameter Setting in Evolutionary Algorithm. London: Springer-Verlag.

51. Michalewicz, Z. 1992. Genetic Algorithms + Data Structure = Evolution Programs. London: Springer-Verlag.

52. Muhlenbein, H. 1992. How Genetic Algorithm Really Work: I. Mutation and HillClimbing. In Manner, R. & Manderick, B. (Eds) Parellel Problem Solving from Nature 2. pp. 15- 25. Belgium: Elsevier.

53. Nelson. R. R. 1995. Recent evolutionary theorizing about economic change. Journal of Economic Literature. 33(1): 48-90.

54. Obitko, M. 1998. Introduction to Genetic Algorithms. http://cs.felk.cvut. cz/ xobitko/ga/. 12 November 2008. Rechenberg, I. 1973.

55. Evolutionsstrategie: Optimierung technischer Systeme nach Prinzipien der biologischen Evolution (Evolution Strategy:

56. Optimization of Technical Systems by Means of Biological Evolution). Stuttgart: Fromman-Holzboog. Roth, A. E. 2002.

57. The economist as engineer: Game theory, experimentation, and computation as tools for design economics. Econometrica, 70(4): 1341-1378.

58. Schaffer, J. D. and Morishima, A. 1987. An Adaptive crossover distribution mechanism for Genetic Algorithms. In Grefensttete, J. J. (Ed) Genetic Algorithms and their Applications: Proceedings of the Second International Conference on Genetic Algorithms. pp. 36-40.

59. Schwefel, H. P. 1977. Numerishce Optimierung von Computer-Modellen mittels der Evolutionsstrategic. Interdisciplinary System Research. 26. Smith, V. 1962.

60. Experimental study of competitive market behavior. Journal Political Economy, 70: 111–137. Spears, W. M. 1995.

61. Adapting Crossover in Evolutionary Algorithm. Proceedings of the Fourth Annual Conference on Evolutionary Programming. pp. 367-384.

62. Cambridge: MIT Press. Tesfatsion, L. 2002. Agent-based computational economics: Growing economies from the bottom up. Artificial Life, 8(1): 55-82.

63. Uckun, S., Bagchi, S. and Kawamura, K. 1993. Managing Genetic Search in Job Shop Scheduling. IEEE Expert: Intelligent Systems and Their Applications, 8(5): 15-24.

64. Wolfstetter, E. 2002. Auctions: An Introduction. Journal of Economic Surveys, 10: 367-420.

Chapter 8

MODELLING THE INNATE IMMUNE SYSTEM

Pedro Rocha, Alexandre Pigozzo, Bárbara Quintela, Gilson Macedo, Rodrigo Santos and Marcelo Lobosco
Federal University of Juiz de Fora, UFJF Brazil

INTRODUCTION

The Human Immune System (His) is a complex network composed of specialized cells, tissues, and organs that is responsible for protecting the organism against diseases caused by distinct pathogenic agents, such as viruses, bacteria and other parasites. The first line of defence against pathogenic agents consists of physical barriers of skin and the mucous membranes. If the pathogenic agents breach this first protection barrier, the innate immune system will be ready for recognize and combat them. The innate immune system is therefore responsible for powerful non-specific defences that prevent or limit infections by most pathogenic microorganisms. The understanding of the innate system is therefore essential, not only because it is the first line of defence of the body, but also because of its quick response. However, its complexity and the intense interaction among several components, make this task extremely complex. Some of its aspects, however, may be better understood if a computational model is used. Modelling and simulation help to understand large complex processes, in particular processes with strongly coupled influences and time-dependent interactions as they occur in the HIS. Also, in silico simulations have the advantage that much less investment in technology, resources and time is needed compared to in vivo experiments, allowing researchers to test a large number of hypotheses in a short period of time. A previous work (Pigozzo et al. (2011)) has developed a mathematical and computational model to simulate the immune response to Lipopolysaccharide (LPS) in a microscopic section of a tissue. The LPS endotoxin is a potent immunostimulant that can induce an acute inflammatory response comparable to that of a bacterial infection. A set of Partial Differential Equations (PDEs) were employed to reproduce the spatial and temporal behaviour of antigens (LPS), neutrophils and cytokines during the first phase of the innate response. Good modelling practices require the evaluation of the confidence in the new proposed model. An important tool used for this purpose is the sensitivity analysis. The sensitivity analysis consists of the study of the impact caused by the variation of input values of a model on the output generated by it. However, this study can be a time consuming task due to the large number of scenarios that must be evaluated. This prohibitive computational cost leads us to develop a parallel version of the sensitivity analysis code using General-purpose Graphics Processing Units (GPGPUs). GPGPUs were chosen because of their ability to process many streams simultaneously. This chapter describes the GPU-based implementation of the sensitivity analysis and also presents some of the sensitivity analysis results. Our experimental results showed that the parallelization was very effective in improving the sensitivity analysis performance, yielding speedups up to 276. The remainder of this chapter is organized as follows. Section 2 includes the background necessary for understanding this chapter. Section 3 describes the mathematical model implemented. Section 4 describes the implementation of the GPU version of the sensitivity analysis. Section 5 presents some of the results of the sensitivity analysis and the speedup obtained. Section 7 presents related works. Our conclusions and plans of future works are presented in Section 8.

BACKGROUND

Biological background

The initial response of the body to an acute biological stress, such as a bacterial infection, is an acute inflammatory response (Janeway et al. (2001)). The strategy of the HIS is to keep some resident macrophages on guard in the tissues to look for any signal of infection. When they find such a signal, the macrophages alert the neutrophils that their help is necessary. The cooperation between macrophages and neutrophils is essential to mount an effective defence, because without the macrophages to recruit the neutrophils to the location of infection, the neutrophils would circulate indefinitely in the blood vessels, impairing the control of huge infections. The LPS endotoxin is a potent immunostimulant that can induce an acute inflammatory response comparable to that of a bacterial infection. After the lyse of the bacteria by the action of cells of the HIS, the LPS can be released in the host, intensifying the inflammatory response and activating some cells of the innate system, such as neutrophils and macrophages. The LPS can trigger an inflammatory response through the interaction with receptors on the surface of some cells.

For example, the macrophages that reside in the tissue recognize a bacterium through the binding of a protein, TLR4, with LPS. The commitment of this receptor activates the macrophage to phagocyte the bacteria, degrading it internally and secreting proteins known as cytokines and chemokines, as well as other molecules. The inflammation of an infectious tissue has many benefits in the control of the infection. Besides recruiting cells and molecules of innate immunity from blood vessels to the location of the infected tissue, it increases the lymph flux containing microorganisms and cells that carry antigens to the neighbours' lymphoid tissues, where these cells will present the antigens to the lymphocytes and will initiate the adaptive response. Once the adaptive response is activated, the inflammation also recruits the effectors cells of the adaptive HIS to the location of infection.

GENERAL-PURPOSE COMPUTATION ON GRAPHICS PROCESSING UNITS - GPGPUS NVIDIA'S COMPUTE UNIFIED DEVICE ARCHITECTURE (CUDA)

NVIDIA (2007)) is perhaps the most popular platform in use for General-Purpose computation on Graphics Processing Units

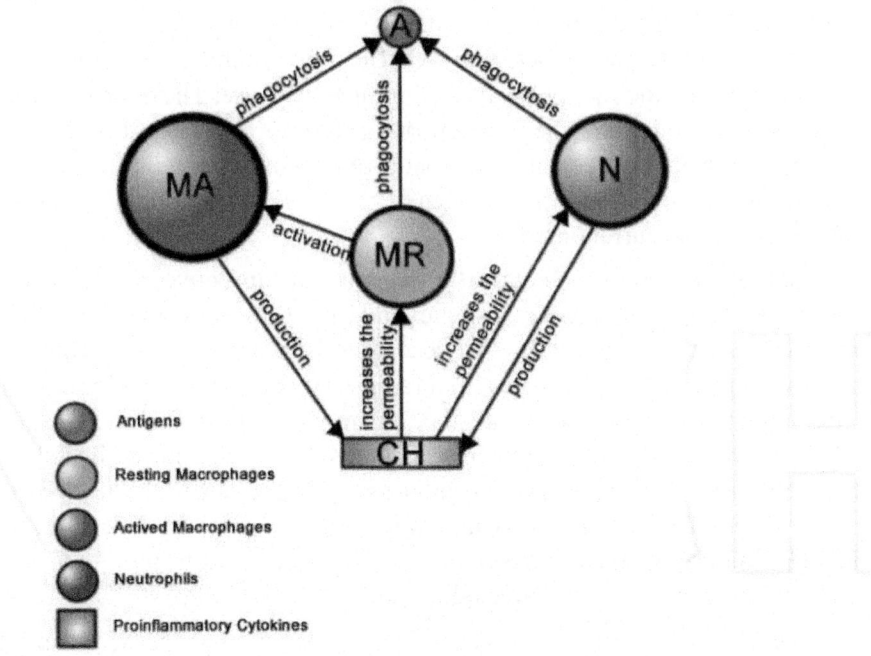

Figure. 1: Relationship between the components

(GPGPUs). CUDA includes C software development tools and libraries to hide the GPGPU hardware from programmers. In CUDA, a parallel function is called kernel. A kernel is a function callable from the CPU and executed on the GPU simultaneously by many threads. Each thread is run by a stream processor. They are grouped into blocks of threads or just blocks. A set of blocks of threads form a grid. When the CPU calls a kernel, it must specify how many threads will be created at runtime. The syntax that specifies the number of threads that will be created to execute a kernel is formally known as the execution conFigureuration, and is flexible to support CUDA's hierarchy of threads, blocks of threads, and grids of blocks. Some steps must be followed to use the GPU: first, the device must be initialized. Then, memory must be allocated in the GPU and data transferred to it. The kernel is then called. After the kernel has finished, results must be copied back to the CPU.

Mathematical model

The model proposed in this chapter is based on a set of Partial Differential Equations (PDEs) originally proposed by Pigozzo et al. (2011). In the original work, a set of PDEs describe the dynamics of the immune response to LPS in a microscopic section of tissue. In particular, the interactions among antigens (LPS molecules), neutrophils and cytokines were modelled. In this chapter, a simplified model of the innate immune system using ODEs is presented to simulate the temporal behaviour of LPS, neutrophils, macrophages and cytokines during the first phase of the immune response. The main differences between our model and the original one (Pigozzo et al. (2011)) are: a) the current model does not consider the spatial dynamics of the cells and molecules and b) the macrophages in two stages of readiness, resting and activated, are introduced in the current model.

Figure 1 presents schematically the relationship between macrophages, neutrophils, proinflammatory cytokines and LPS. LPS cause a response in both macrophages and neutrophils, that recognize LPS and phagocyte them. The process of phagocytosis induces, in a rapid way, the apoptosis of neutrophils. This induction is associated with the generation of reactive oxygen species (ROS) (Zhang et al. (2003)). The resting macrophages become activated when they find LPS in the tissue. The pro-inflammatory cytokine is produced by both active macrophages and neutrophils after they recognize LPS. It induces an increase in the endothelial permeability allowing more neutrophils to leave the blood vessels and enter the infected tissue.

Our set of equations is given below, where RM, AM, A, N and CH represent the population of resting macrophages, activated macrophages, LPS, neutrophils and pro-inflammatory cytokines, respectively. The dynamics of LPS is modelled with Equation 1.

$$\begin{cases} \frac{dA}{dt} = -\mu_A A - (\lambda_{N|A}.N + \lambda_{AM|A}.AM + \lambda_{RM|A}.RM).A \\ A(0) = 20 \end{cases} \tag{1}$$

The term $\mu_A A$ models the decay of LPS, where μ_A is its decay rate.

The term $-(\lambda_{N|A}.N + \lambda_{AM|A}.AM + \lambda_{RM|A}.RM)$

A models the phagocytosis of LPS by macrophages and neutrophils, where

$\lambda_{N|A}$ is the phagocytosis rate of neutrophils, $\lambda_{N|A}$ is the phagocytosis

rate of active macrophages, and $\lambda_{AM|A}$ is the phagocytosis rate of resting macrophages. Neutrophils are modelled with Equation 2.

$$\begin{cases} permeability_N = (P_N^{max} - P_N^{min}).\frac{CH}{CH+keqch} + P_N^{min} \\ source_N = permeability_N.(N^{max} - N) \\ \frac{dN}{dt} = -\mu_N N - \lambda_{A|N}A.N + source_N \\ N(0) = 0 \end{cases} \tag{2}$$

The term permeability^N uses a Hill equation (Goutelle et al. (2008)) to model how permeability of the endothelium of the blood vessels depends on the local concentration of cytokines. Hill equations are also used, for example, to model drug dose-response relationships (Wagner (1968)). The idea is to model the increase in the permeability of the endothelium according to the concentration of the pro-inflammatory cytokines into the endothelium. In the Hill equation, P_N^{max} represents the maximum rate of increase of endothelium permeability to neutrophils induced by pro-inflammatory cytokines, P_N^{min}. represents the minimum rate of increase of endothelium permeability induced by pro-inflammatory cytokines and keqch is the concentration of the pro-inflammatory cytokine that exerts 50% of the maximum effect in the increase of the permeability. The term $\mu_N N$ models the neutrophil apoptosis, where μ_N is the rate of apoptosis. The term $\lambda_{A|N}A.N$ models the neutrophil apoptosis induced by the phagocytosis, where $\lambda_{A|N}$ represent the rate of this induced apoptosis. The term sourceN represents the source term of neutrophil, that is, the number of neutrophils that is entering the tissue from the blood vessels. This number depends on the endothelium permeability (permeability$_N$) and the capacity of the tissue to support the entrance of neutrophils (N^{max}), that can also represent the blood concentration of Neutrophils. The dynamics of cytokine is presented in Equation 3.

$$\begin{cases} \frac{dCH}{dt} = -\mu_{CH}CH + (\beta_{CH|N}N + \beta_{CH|AM}AM).A.(1 - \frac{CH}{chInf}) \\ CH(0) = 0 \end{cases}$$

$$(3)$$

The term $\mu_{CH}CH$ models the pro-inflammatory cytokine decay, where μCH is the decay rate. The term $(\beta_{CH|N}N + \beta_{CH|AM}AM).A$ models the production of the pro-inflammatory cytokine by the neutrophils and activated macrophages, where $\beta_{CH|N}$ and $\beta_{CH|AM}$ are the rate of this production by neutrophils and macrophages, respectively.

Equation 4 presents the dynamics of the resting macrophages.

$$\begin{cases} permeability_{RM} = (P_{RM}^{max} - P_{RM}^{min}).\frac{CH}{CH+keqch} + P_{RM}^{min} \\ source_{RM} = permeability_{RM}.(M^{max} - (RM + AM)) \\ \frac{dRM}{dt} = -\mu_{RM}RM - \lambda_{RM|A}.RM.A + source_{RM} \\ RM(0) = 1 \end{cases}$$

$$(4)$$

The term $permeability_{RM}$ models how permeability of the endothelium of the blood vessels to macrophages depends on the local concentration of cytokines. The term $\mu_{RM}RM$ models the resting macrophage apoptosis, where μ_{RM} is the rate of apoptosis. Finally, the dynamics of activate macrophages is presented in Equation 5.

$$\begin{cases} \frac{dAM}{dt} = -\mu_{AM}AM + \lambda_{RM|A}.RM.A \\ AM(0) = 0 \end{cases}$$

$$(5)$$

The term $\mu_{AM}RM$ models the activated macrophage apoptosis, where μRM is the rate of apoptosis.

IMPLEMENTATION

The sensitivity analysis consists in the analysis of impacts caused by variations of parameters and initial conditions of the mathematical model against its dependent variables (Saltelli et al. (2008)). If a parameter causes a drastic change in the output of the problem, after suffering a minor change in its initial value, it is thought that this parameter is sensitive to the problem studied. Otherwise, this variable has little impact in the model. The sensitivity analysis is used to improve the understanding of the mathematical model as it allows us to identify input parameters that are more relevant for the model, i.e. the values of these parameters should be carefully estimated. In this chapter we use a brute force approach to exam the influence of the 19 parameters present in the equation and two of the initial conditions. A small change in the value of each parameter is done, and then the model is solved again for this new parameter set. This process is done many times, since all combinations of distinct values of parameters and initial conditions must be considered. We analyse the impact of changing one coefficient at a time. The parameters and initial conditions were adjusted from -100% to + 100% (in steps of 2%) of their initial values, except for some parameters, that were also adjusted from -100% to + 100%, but in steps of 20%. The combination of all different set of parameters and initial conditions give us a total of 450,000 system of ODEs that must be evaluated in this work. The sequential code that implements the sensitivity analysis was first implemented in C. Then the code was parallelized using CUDA. The parallel code is based on the idea that each combination of distinct values of parameters and initial conditions can be computed independently by a distinct CUDA thread. The number of threads that will be used during computation depends on the GPU characteristics. In particular, the number of blocks and threads per block are chosen taking into account two distinct values defined by the hardware: a) the warp size and b) the maximum number of threads per block. The forward Euler method was used for the numerical solution of the systems of ODEs with a time-step of 0.0001 days. The models were simulated to represent a total period equivalent to 5 days after the initial infection.

EXPERIMENTAL EVALUATION

In this section the experimental results obtained by the execution of both versions of our simulator of the innate system, sequential and parallel, are presented. The experiments were performed on a 2.8 GHz Intel Core i7-860 processor, with 8 GB RAM, 32 KB L1 data cache, 8 MB L2 cache with a NVIDIA GeForce 285 GTX. The system runs a 64-bits version of Linux kernel 2.6.31 and version 3.0 of CUDA toolkit. The gcc version 4.4.2 was used to compile all versions of our code. The NVIDIA GeForce 285 GTX has 240 stream processors, 30 multiprocessors, each one with 16KB of shared memory, and 1GB of global memory. The number of threads per block are equal to 879, and each block has 512 threads. The codes were executed 3 times to all versions of our simulator, and the average execution time for each version of the code is presented in Table 1. The standard deviation obtained was negligible. The execution times were used to calculate the speedup factor. The speedup were obtained by dividing the sequential execution time of the simulator by its parallel version.

Table 1: Serial and parallel execution times

Sequential	285 GTX	Speedup Factor
4,315.47s	15.63s	276.12

All times are in seconds. The results reveal that our CUDA version was responsible for a significant improvement in performance: a speedup of 276 was obtained. This expressive gain was due to the embarrassingly parallel nature of computation that must be performed. In particular, the same computation must be performed for a huge amount of data, and there are no dependency and/or communication between parallel tasks.

Simulation

To study the importance of some cells, molecules and processes in the dynamics of the innate immune response, a set of simulations were performed for distinct values of parameters and initial conditions. Table 2 presents the initial conditions and the values of the parameters used in the simulations of all cases. Exceptions to the values presented in Table 2 are highlighted in the text.

The complete set of equations that has been simulated, including the initial values used, are presented by Equation 6:

$$
\begin{cases}
\frac{dA}{dt} = -\mu_A A - (\lambda_{N|A}.N + \lambda_{AM|A}.AM + \lambda_{RM|A}.RM).A \\
A(0) = 20|40 \\
\\
permeability_N = (P_N^{max} - P_N^{min}).\frac{CH}{CH+keqch} + P_N^{min} \\
source_N = permeability_N.(N^{max} - N) \\
\frac{dN}{dt} = -\mu_N N - \lambda_{A|N}A.N + source_N \\
N(0) = 0 \\
\\
permeability_{RM} = (P_{RM}^{max} - P_{RM}^{min}).\frac{CH}{CH+keqch} + P_{RM}^{min} \\
source_{RM} = permeability_{RM}.(M^{max} - (RM + AM)) \\
\frac{dRM}{dt} = -\mu_{RM} RM - \lambda_{RM|A}.RM.A + source_{RM} \\
RM(0) = 1 \\
\\
\frac{dAM}{dt} = -\mu_{AM} AM + \lambda_{RM|A}.RM.A \\
AM(0) = 0 \\
\\
\frac{dCH}{dt} = -\mu_{CH} CH + (\beta_{CH|N}N + \beta_{CH|AM}AM).A.(1 - \frac{CH}{chInf}) \\
CH(0) = 0
\end{cases}
$$

$$(6)$$

It should be noticed that in this case two distinct initial values for A(0) will be used: A(0) = 20 and A(0) = 40. The sensitivity analysis has shown that two parameters are relevant to the model: the capacity of the tissue to support the entrance of new neutrophils (N^{max}) and the phagocytosis rate of LPS by neutrophils $(\lambda_{N|A})$.

N^{max} is the most sensitive parameter in the model. The capacity of the tissue to support the entrance of new neutrophils is directed related to the permeability of the endothelial cells, which form the linings of the blood vessels. If a positive adjustment is made in the parameter related to the permeability, then there are more neutrophils entering into the tissue. This larger amount of neutrophils into the tissue has many consequences: first, more cells are phagocyting, so the amount of LPS reduces faster. Second, a smaller amount of resting macrophages becomes active, because there is less LPS into the tissue. Third, a larger amount of cytokines are produced, since neutrophils are the main responsible for this production. If a negative adjustment is made, the inverse effect can be observed: with a smaller amount of neutrophils in the tissue, more resting macrophages become active. Also, a smaller amount of cytokines are produced. Figureures 2 to 6 illustrate this situation. It can be observed that the LPS decays faster when Nmax achieves its maximum value.

Table 2: Initial conditions, parameters and units

Parameter	Value	Unit	Reference
N_0	0	cell	estimated
CH_0	0	cell	estimated
A_0	20	cell	estimated
RM_0	1	cell	estimated
AM_0	0	cell	estimated
μ_{CH}	7	$1/day$	estimated
μ_N	3.43	$1/day$	estimated
μ_A	0	$1/day$	Su et al. (2009)
μ_{RM}	0.033	$1/day$	Su et al. (2009)
μ_{AM}	0.07	$1/day$	Su et al. (2009)
$\lambda_{N|A}$	0.55	$\frac{1}{cell.day}$	Su et al. (2009)
$\lambda_{A|N}$	0.55	$\frac{1}{cell.day}$	Su et al. (2009)
$\lambda_{AM|A}$	0.8	$\frac{1}{cell.day}$	Su et al. (2009)
$\beta_{CH|N}$	1	$\frac{1}{cell.day}$	estimated
$\beta_{CH|AM}$	0.8	$\frac{1}{cell.day}$	estimated
N^{max}	8	$cell$	estimated
MR^{max}	6	$cell$	estimated
P_N^{max}	11.4	$\frac{1}{day}$	based on Price et al. (1994)
P_N^{min}	0.0001	$\frac{1}{day}$	estimated
P_{RM}^{max}	0.1	$\frac{1}{day}$	estimated
P_{RM}^{min}	0.01	$\frac{1}{day}$	estimated
$chInf$	3.6	$cell$	based on de Waal Malefyt et al. (1991)
$keqch$	1	$cell$	estimated
$\lambda_{RM|A}$	0.1	$\frac{1}{cell.day}$	estimated

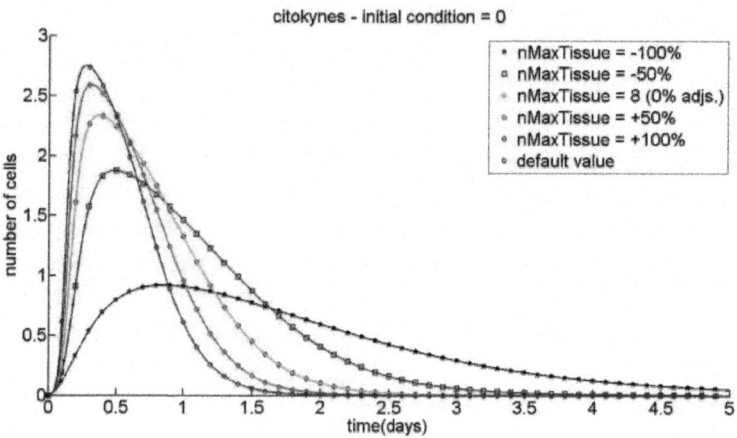

Figure. 2: Temporal evolution of cytokines with A(0) = 20 and for distinct values of N^{max} .

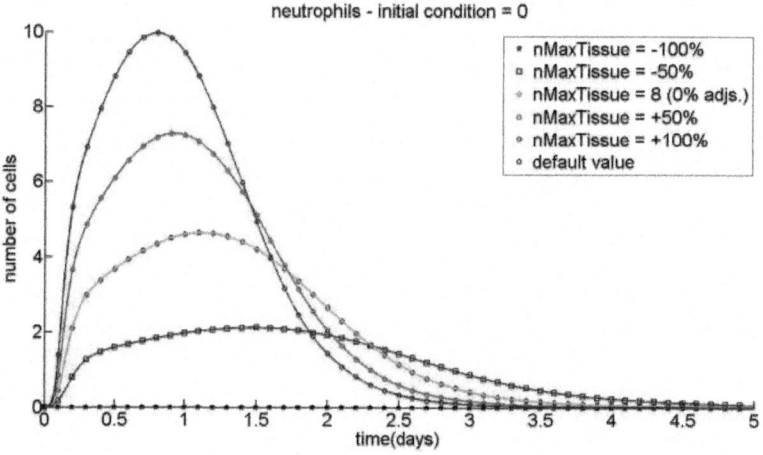

Figure. 3: Temporal evolution of neutrophils with A(0) = 20 and for distinct values of Nmax .

In the second scenario, with the double of LPS and starting with just one resting macrophage, it can be observed that bringing more neutrophils into the tissue do not reduce the number of resting macrophages that become active. This happens due to the larger amount of LPS in this scenario when compared to the previous one. The larger amount of activated macrophages also explains why the amount of cytokines in this scenario is larger than in the previous one. Figureures 7 to 11 present the complete scenario.

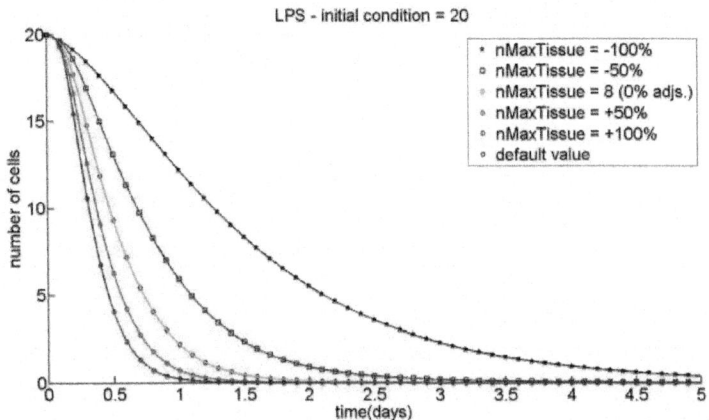

Figure. 4: Temporal evolution of LPS with $A(0) = 20$ and for distinct values of Nmax

The third scenario presents the results obtained when the initial amount of LPS is again equal to 20. This scenario revels that the second most sensitive parameter is $\lambda_{N|A}$. $\lambda_{N|A}$ is responsible for determining how effective is the phagocitosis of the neutrophils in tissue. It can be observed in Figureures 12 to 16 that a negative adjustment in this tax makes the neutrophil response to be less effective against LPS, while a positive adjustment in the tax makes the neutrophil response to be more effective. Resting macrophages and activated macrophages are also affected by distinct values of $\lambda_{N|A}$. Increasing the value of $\lambda_{N|A}$ causes the neutrophils

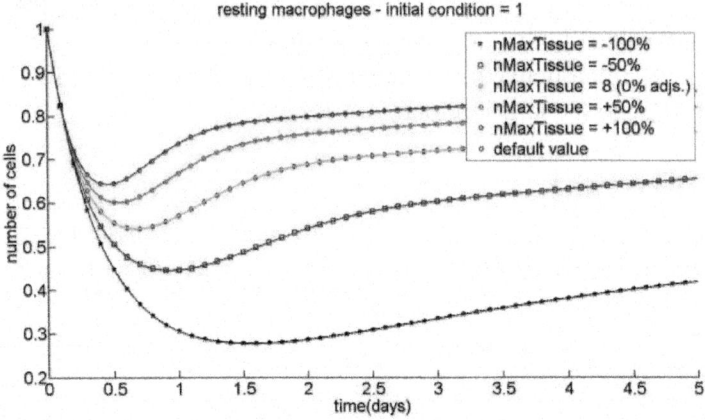

Figure. 5: Temporal evolution of resting macrophages with $A(0) = 20$ and for distinct values of N^{max}

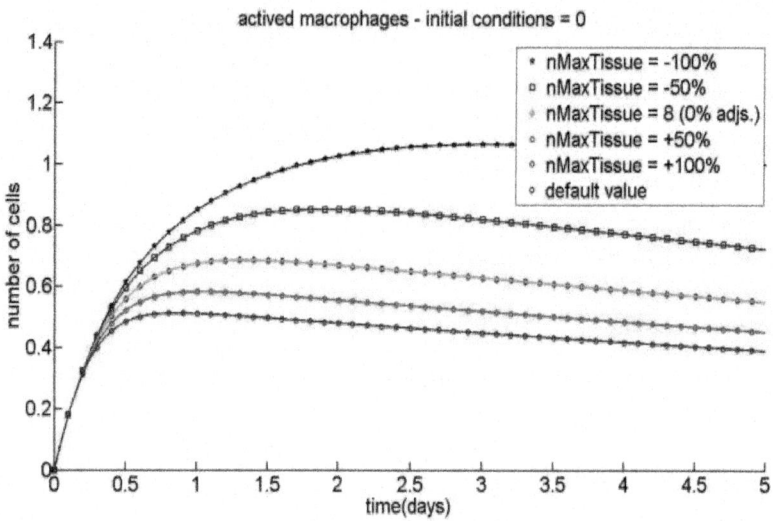

Figure. 6: Temporal evolution of activate macrophages with $A(0) = 20$ and for distinct values of Nmax

to produced more cytokines, so more macrophages can migrate into the tissue through blood vessel, and also there are more cells into the tissue that can phagocyte LPS. The last scenario is presented by Figureures 17 to 21. In this scenario, the amount of LPS is doubled when compared to the previous one. It can be observed that distinct values used as initial conditions for LPS only changes how long it takes to the complete elimination of LPS. It can also be observed that both macrophages populations are affected by the larger amount of LPS. In particular, the amount of macrophages is slightly higher in this scenario due to the larger amount of LPS.

RELATED WORKS

This section presents some models and simulators of the HIS found in the literature. Basically two distinct approaches are used: ODEs and PDEs.

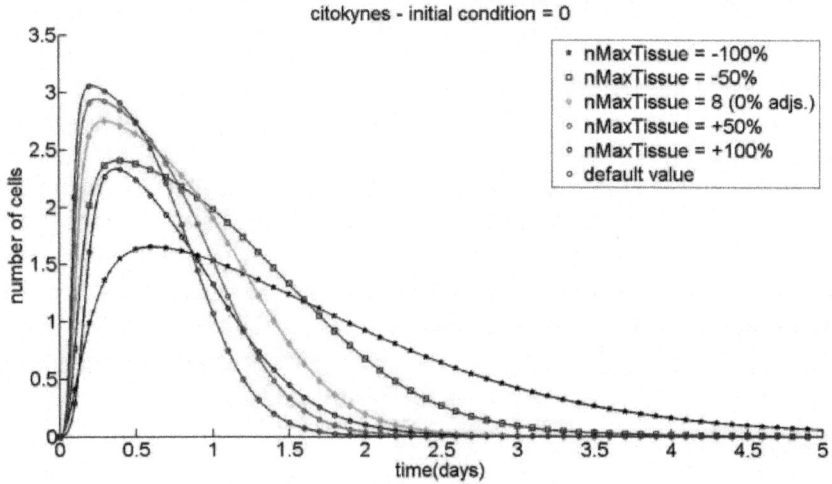

Figure. 7: Temporal evolution of cytokines with A(0) = 40 and for distinct values of N^{max}.

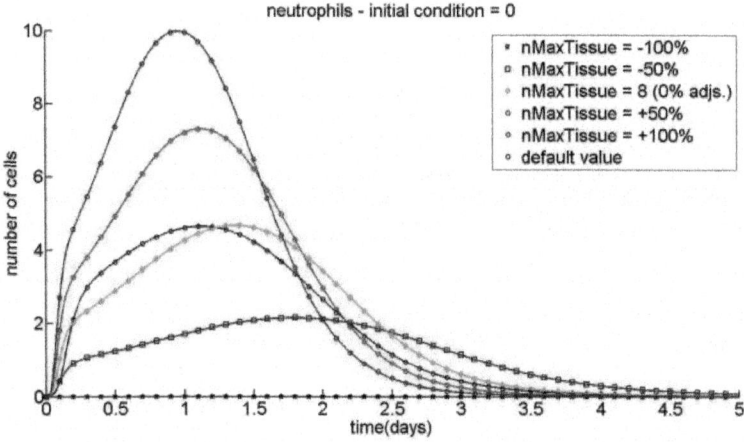

Figure. 8. Temporal evolution of neutrophils with A(0) = 40 and for distinct values of N^{max}.

ODES MODELS

A model of inflammation composed by ODEs in a three-dimensional domain considering three types of cells/molecules has been proposed by Kumar et al. (2004): the pathogen and two inflammatory mediators. The model was able to reproduce some experimental results depending on the values used for initial conditions and parameters. The authors described the results of the sensitivity analysis and some therapeutic strategies were suggested from this analysis. The work was then extended (Reynolds et al. (2006)) to investigate the advantages of an anti-inflammatory response dependent on time. In this extension, the mathematical model was built from simpler models, called reduced models. The mathematical model (Reynolds et al. (2006)) consists of a system of ODEs with four equations to model: a) the pathogen; b) the active phagocytes; c) tissue damage; and d) anti-inflammatory mediators. A new adaptation of the first model (Kumar et al. (2004)) was proposed to simulate many scenarios involving repeated doses of endotoxin (Day et al. (2006)). In this work the results obtained through experiments with mouse are used to guide the in silico experiments seeking to recreate these results qualitatively.

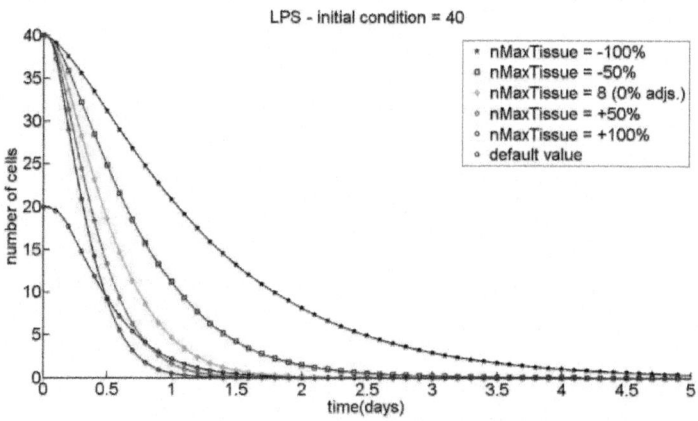

Figure. 9: Temporal evolution of LPS with A(0) = 40 and for distinct values of Nmax

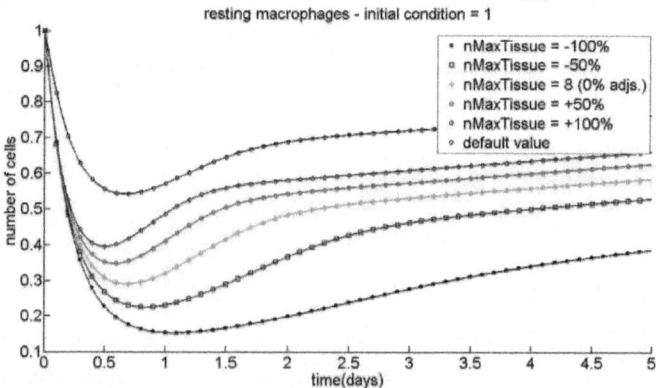

Figure. 10: Temporal evolution of resting macrophages with A(0) = 40 and for distinct values of Nmax .

A one-dimensional model to show if and when leukocytes successfully defend the body against a bacterial infection is presented in Keener & Sneyd (1998). A phase-plane method is then used to study the influence of two parameters, the enhanced leukocyte emigration from bloodstream and the chemotactic response of the leukocytes to the attractant. Finally, one last work (Vodovotz et al. (2006)) developed a more complete system of ODEs of acute inflammation, including macrophages, neutrophils, dendritic cells, Th1 cells, the blood pressure, tissue trauma, effector elements such as $iNOS$, NO_2^- and NO_3^-, pro-inflammatory and anti-inflammatory cytokines, and coagulation factors. The model has proven to be useful in simulating the inflammatory response induced in mice by endotoxin, trauma and surgery or surgical bleeding, being able to predict to some extent the levels of TNF, IL-10, IL-6 and reactive products of NO ($\dot{NO}_2^-$ and $\dot{NO}_3^-$).

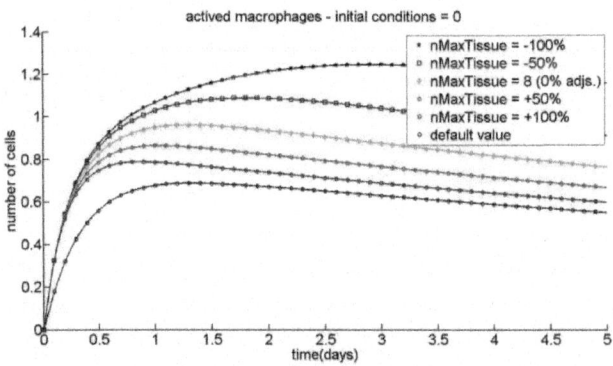

Figure. 11: Temporal evolution of activate macrophages with $A(0) = 40$ and for distinct values of N^{max} .

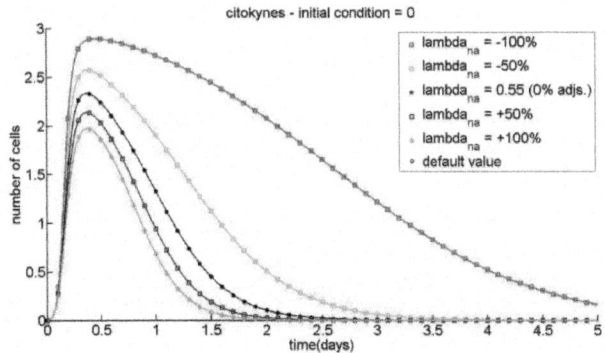

Figure. 12: Temporal evolution of cytokines with $A(0) = 20$ and for distinct values of $\lambda_{N|A}$.

PDES MODELS

The model proposed by Su et al. (2009) uses a system of partial differential equations (PDEs) to model not only the functioning of the innate immune system, as well as the adaptive immune system. The model considers the simplest form of antigen, the molecular constituents of pathogens patterns, taking into account all the basic factors of an immune response: antigen, cells of the immune system, cytokines and chemokines. This model captures the following stages of the immune response: recognition, initiation, effector response and resolution of infection or change to a new equilibrium state (steady state). The model can reproduce important phenomena of the HIS such as a) temporal order of arrival of cells at the site of infection, b) antigen presentation by dendritic cells, macrophages to regulatory T cells d) production of pro-inflammatory and anti-inflammatory cytokines and e) the phenomenon of chemotaxis.

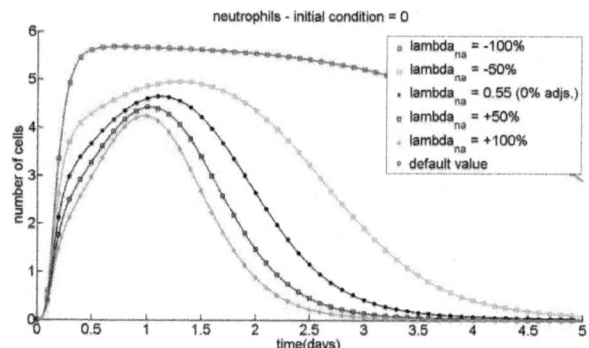

Figure. 13: Temporal evolution of neutrophils with A(0) = 20 and for distinct values of $\lambda_{N|A}$.

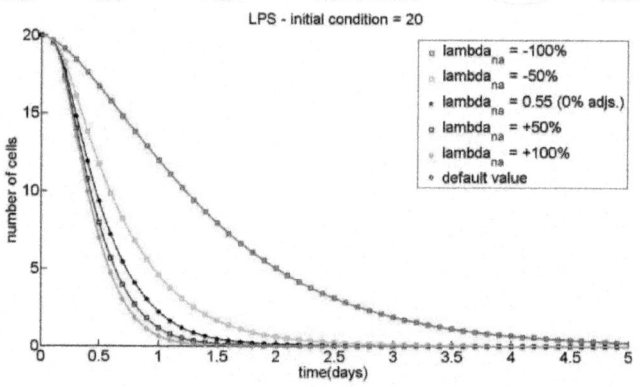

Figure. 14: Temporal evolution of LPS with A(0) = 20 and for distinct values of $\lambda_{N|A}$.

Pigozzo et al. (2011) present a PDE model to simulate the immune response to lipopolysaccharide (LPS) in a microscopic section of a tissue, reproducing, for this purpose, the initiation, maintenance and resolution of immune response.

Other works

Several proposals which attempt to model both the innate and the adaptive HIS can be found in the literature. An ODE model is used to describe the interaction of HIV and tuberculosis with the immune system (Denise & Kirschner (1999)). Other work focus on models of HIV and T-lymphocyte dynamics, and includes more limited discussions of hepatitis C virus (HCV), hepatitis B virus (HBV), cytomegalovirus (CMV) and lymphocytic choriomeningitis virus (LCMV) dynamics and interactions with the immune system (Perelson (2002)). An ODE model of cell-free viral spread of HIV in a compartment was proposed by Perelson et al. (1993). Another interesting work tries to integrate the immune system in the general physiology of the host and considers the interaction between the immune and neuroendocrine system

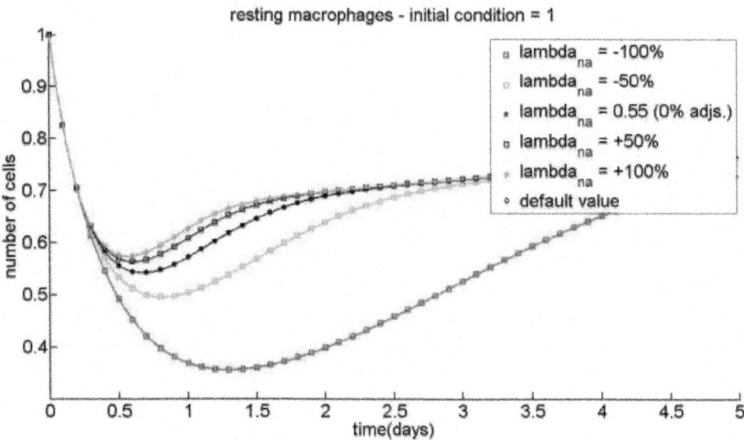

Figure. 15: Temporal evolution of resting macrophages with $A(0) = 20$ and for distinct values of $\lambda_{N|A}$.

(Muraille et al. (1996)). Klein (1980) presents and compares three mathematical models of B cell differentiation and proliferation. ImmSim (Bezzi et al. (1997); Celada & Seiden (1992)) is a simulator of the HIS that implements the following mechanisms: immunological memory, affinity maturation, effects of hypermutation, autoimmune response, among others. CAFISS (a Complex Adaptive Framework for Immune System Simulation)

(Tay & Jhavar (2005)) is a framework used for modelling the immune system, particularly HIV attack. SIMMUNE (Meier-Schellersheim & Mack (1999)) allows users to model cell biological systems based on data that describes cellular behaviour on distinct scales. Although it was developed to simulate immunological phenomena, it can be used in distinct domains. A similar tool is CyCells (Warrender (2004)), designed to study intercellular relationships.

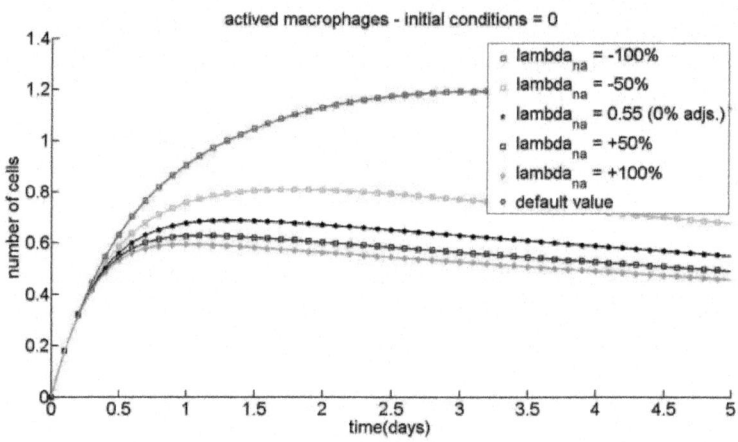

Figure. 16: Temporal evolution of activate macrophages with A(0) = 20 and for distinct values of $\lambda_{N|A}$

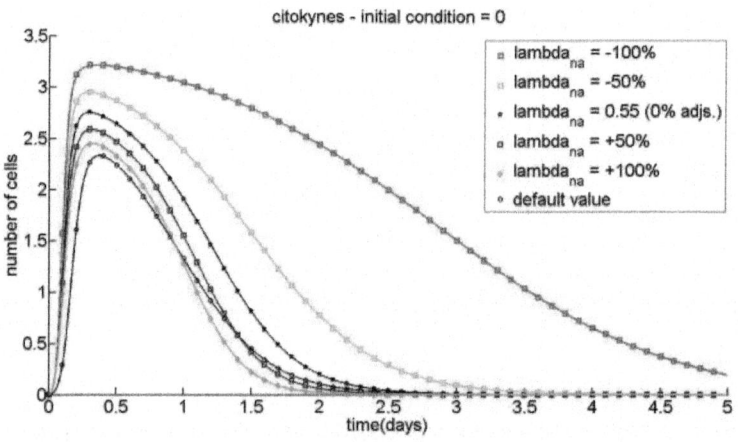

Figure. 17: Temporal evolution of cytokines with A(0) = 40 for distinct values of $\lambda_{N|A}$.

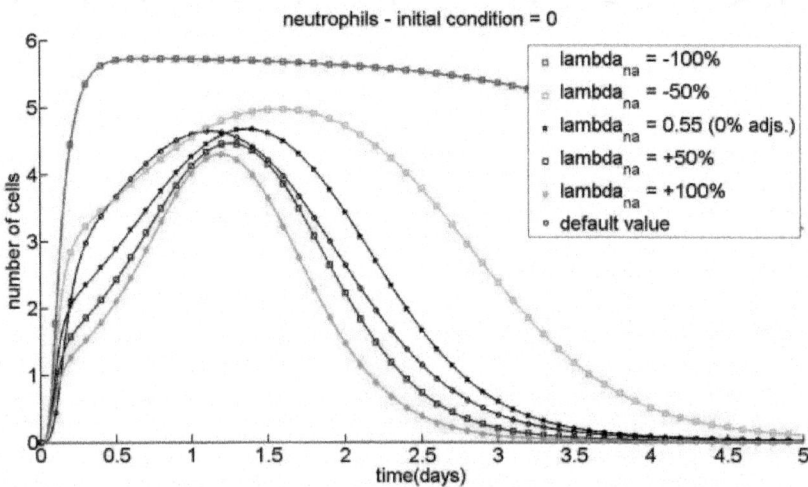

Figure. 18: Temporal evolution of neutrophils with A(0) = 40 for distinct values of $\lambda_{N|A}$.

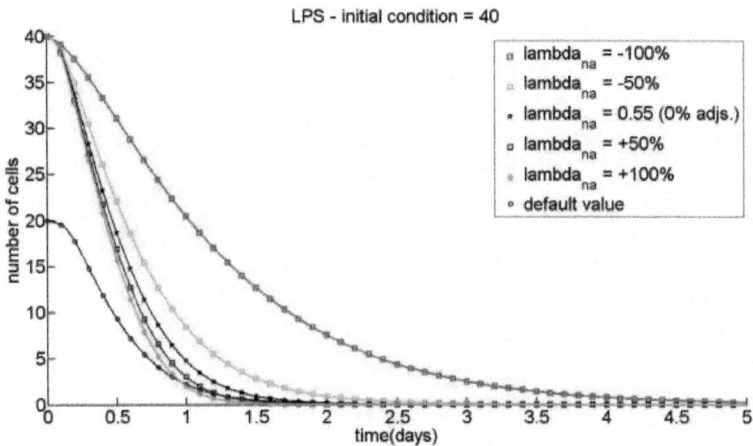

Figure. 19: Temporal evolution of LPS with A(0) = 40 for distinct values of $\lambda_{N|A}$.

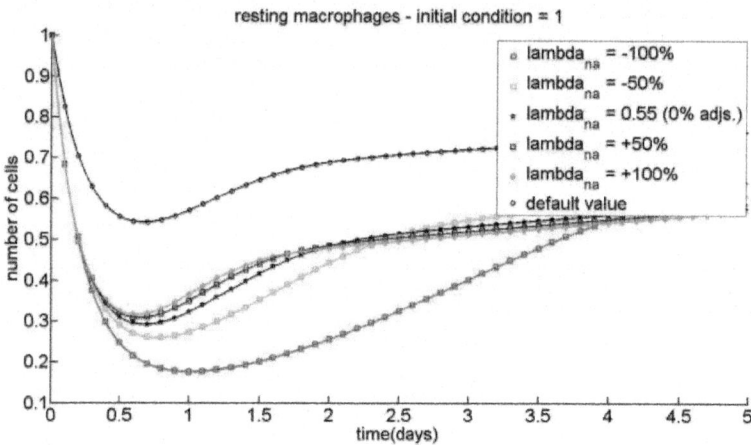

Figure. 20: Temporal evolution of resting macrophages with $A(0) = 40$ for distinct values of $\lambda_{N|A}$.

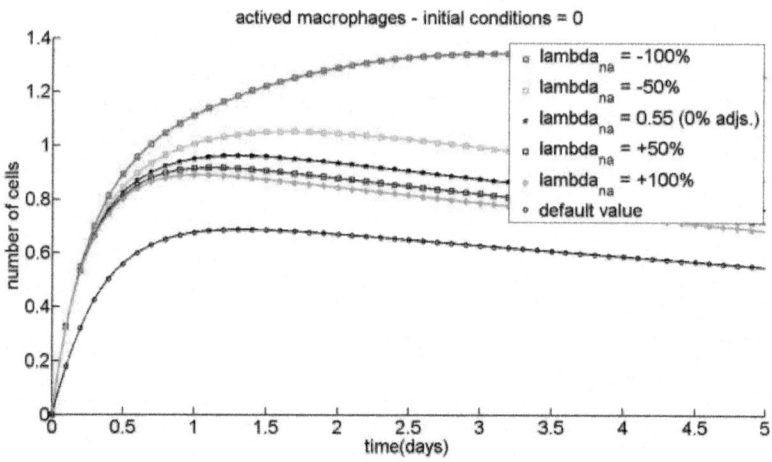

Figure. 21: Temporal evolution of activate macrophages with $A(0) = 40$ for distinct values of $\lambda_{N|A}$.

CONCLUSION AND FUTURE WORKS

In this chapter we presented the sensitivity analysis of a mathematical model that simulates the immune response to LPS in a microscopic section of a tissue. The results have shown that the two most relevant parameters of the model are: the capacity of the tissue to support the entrance of more neutrophils and the phagocytosis rate of LPS by neutrophils. The sensitivity analysis can be a time consuming task due to the large number of scenarios that must be evaluated. This prohibitive computational cost leads us to develop a parallel version of the sensitivity analysis code using GPGPUs. Our experimental results showed that the parallelization was very effective in improving the sensitivity analysis performance, yielding speedups up to 276.

As future works, we plan to implement a more complete mathematical model including, for example, new cells (Natural Killer, dendritic cells and the complement system), others proinflammatory cytokines, anti-inflammatory cytokine, molecules and others processes involved in the immune responses.

ACKNOWLEDGEMENT

The authors would like to thank FAPEMIG, CNPq (479201/2010-2), CAPES and UFJF for supporting this study

REFERENCES

1.   Bezzi, M., Celada, F., Ruffo, S. & Seiden, P. E. (1997). The transition between immune and.disease states in a cellular automaton model of clonal immune response, Physica A:.Statistical and Theoretical Physics 245(1-2): 145 – 163..URL: http://www.sciencedirect.com/science/article/ B6TVG-3W34FV4-8/2/992f79c98f0d0e.31f1bb36b3f524426d

2.   Celada, F. & Seiden, P. E. (1992). A computer model of cellular interactions in the immune.system, Immunology Today 13(2): 56 – 62.. URL: http://www.sciencedirect.com/science/article/B6VHW-4805SYB-P/2/074bd180cec58.7021d6ed7b96be84125

3.   Day, J., Rubin, J., Vodovotz, Y., Chow, C. C., Reynolds, A. & Clermont, G. (2006). A reduced.mathematical model of the acute inflammatory response ii. capturing scenarios of.repeated endotoxin administration., J Theor Biol 242(1): 237–256..URL: http://dx.doi.org/10.1016/j. jtbi.2006.02.015

4.   de Waal Malefyt, R., Abrams, J., Bennett, B., Figuredor, C. & de Vries, J. (1991). Interleukin.10(il-10) inhibits cytokine synthesis by human monocytes: an autoregulatory role.of il-10 produced by monocytes., J Exp Med 174(5): 1209–20–..URL: http://ukpmc.ac.uk/abstract/ MED/1940799

5.   Denise & Kirschner (1999). Dynamics of co-infection with m. tuberculosis and hiv-1, Theoretical.Population Biology 55(1): 94 – 109..URL: http:// www.sciencedirect.com/science/article/pii/S004058099891382X

6. Goutelle, S., Maurin, M., Rougier, F., Barbaut, X., Bourguignon, L., Ducher, M. & Maire, P..(2008). The hill equation: a review of its capabilities in pharmacological modelling,.Fundamental & clinical pharmacology 22(6): 633–648..URL: http://dx.doi.org/10.1111/j.1472-8206.2008.00633.x

7. Janeway, C., Murphy, K. P., Travers, P., Walport, M. & Janeway, C. (2001). Immunobiology, 5th.ed. edn, Garland Science, New York and London.. Keener, J. & Sneyd, J. (1998). Mathematical physiology, Springer-Verlag New York, Inc., New.York, NY, USA.

8. Klein, P. (1980). Mathematical models of antibody response, Folia Microbiologica 25: 430–438..10.1007/BF02876697..URL: http://dx.doi.org/10.1007/BF02876697

9. Kumar, R., Clermont, G., Vodovotz, Y. & Chow, C. C. (2004). The dynamics of acute.inflammation, Journal of Theoretical Biology 230(2): 145–155.

10. .URL:http://www.sciencedirect.com/science/article/B6WMD-4D1TSCK-2/2/44a01fc313cd.567f0861e5b6c36fc80f

11. Meier-Schellersheim, M. & Mack, G. (1999). Simmune, a tool for simulating and analyzing.immune system behavior..URL: http://www.citebase.org/abstract?id=oai:arXiv.org:cs/9903017

12. Muraille, E., Thieffry, D., Leo, O. & Kaufman, M. (1996). Toxicity and neuroendocrine.regulation of the immune response: A model analysis, Journal of Theoretical Biology.183(3): 285 – 305..URL: http://www.sciencedirect.com/science/article/pii/S0022519396902210

13. NVIDIA (2007). Nvidia cuda programming guide, Technical report, NVIDIA Corporation..Perelson, A. S. (2002). Modelling viral and immune system dynamics., Nat Rev Immunol.2(1): 28–36..URL: http://dx.doi.org/10.1038/nri700

14. Perelson, A. S., Kirschner, D. E. & de Boer, R. (1993). Dynamics of hiv infection of cd4+ t cells,.Mathematical Biosciences 114(1): 81 – 125.

15. Pigozzo, A. B., Macedo, G. C., dos Santos, R. W. & Lobosco, M. (2011). Implementation of a.computational model of the innate immune system, ICARIS, pp. 95–107.

16. Price, T., Ochs, H., Gershoni-Baruch, R., Harlan, J. & Etzioni, A. (1994). In vivo neutrophil and.lymphocyte function studies in a patient with leukocyte adhesion deficiency type ii,.Blood 84(5): 1635–1639..

URL: http://bloodjournal.hematologylibrary.org/cgi/content/abstract/
bloodjournal;84/5/1635

17. Reynolds, A., Rubin, J., Clermont, G., Day, J., Vodovotz, Y. & Ermentrout,
     G. B. (2006). A.reduced mathematical model of the acute inflammatory
     response: I. derivation of.model and analysis of anti-inflammation,
     Journal of Theoretical Biology 242(1): 220–236..URL: http://www.
     sciencedirect.com/science/article/B6WMD-4JMKWTP-2/2/5ae6086e6a
     0.80ecb9bfa17c6f2a947c9

18. Saltelli, A., Ratto, M., Andres, T., Campolongo, F., Cariboni, J., Gatelli,
     D., Saisana, M. &.Tarantola, S. (2008). Global Sensitivity Analysis: The
     Primer, 1 edn, Wiley..Su, B., Zhou, W., Dorman, K. S. & Jones, D. E. (2009).
     Mathematical modelling of.immune response in tissues, Computational
     and Mathematical Methods in Medicine:.An Interdisciplinary Journal of
     Mathematical, Theoretical and Clinical Aspects of Medicine.10: 1748–
     6718.

19. Tay, J. C. & Jhavar, A. (2005). Cafiss: a complex adaptive framework for
     immune system.simulation, Proceedings of the 2005 ACM symposium
     on Applied computing, SAC '05,.ACM, New York, NY, USA, pp. 158–
     164..URL: http://doi.acm.org/10.1145/1066677.1066716

20. Vodovotz, Y., Chow, C. C., Bartels, J., Lagoa, C., Prince, J. M., Levy, R.
     M., Kumar, R., Day, J.,.Rubin, J., Constantine, G., Billiar, T. R., Fink, M.
     P. & Gilles Clermont, K. (2006). In.silico models of acute inflammation
     in animals..Wagner, J. G. (1968). Kinetics of pharmacologic response
     i. proposed relationships between.response and drug concentration in
     the intact animal and man, Journal of Theoretical.Biology 20(2): 173
     – 201..URL: http://www.sciencedirect.com/science/article/B6WMD-
     4F1Y9M7-N2/2/9bf7ec729de.0947563c9645c61399a34

21. Warrender, C. E. (2004). Modeling intercellular interactions in the
     peripheral immune system, PhD thesis, Albuquerque, NM, USA.
     AAI3156711

22. Zhang, B., Hirahashi, J., Cullere, X. & Mayadas, T. N. (2003).
     Elucidation of molecular events leading to neutrophil apoptosis following
     phagocytosis, The Journal of biological chemistry 278: 28443–28454

Chapter 9

OPTIMAL DESIGN OF POWER SYSTEM CONTROLLER USING BREEDER GENETIC ALGORITHM

K. A. Folly and S. P. Sheetekela

University of Cape Town Private Bag., Rondebosch 7701 South Africa

INTRODUCTION

Genetic Algorithms (GAs) have recently found extensive applications in solving global optimization problems (Mitchell, 1996). GAs are search algorithms that use models based on natural biological evolution (Goldberg, 1989). They are intrinsically robust search and optimization mechanisms and offer several advantages over traditional optimization techniques, including the ability to effectively search large space without being caught in local optimum. GAs do not require the objective function to have properties such as continuity or smoothness and make no use of hessians or gradient estimates. In the last few years, Genetic Algorithms (GAs) have shown their potentials in many fields, including in the field of electrical power systems. Although GAs provide robust and powerful adaptive search mechanism, they have several drawbacks (Mitchell, 1996). Some of these drawbacks include the problem of "genetic drift" which prevents GAs from maintaining diversity in its population. Once the population has converged, the crossover operator becomes ineffective in exploring new portions of the search space. Another drawback is the difficulty to optimize the GAs' operators (such as population size, crossover and mutation rates) one at a time. These operators (or parameters) interact with one another in a nonlinear manner. In particular, optimal population size, crossover rate, and mutation rate are likely to change over the course of a single run (Baluja, 1994). From the user's point of view, the selection of GAs' parameters is not a trivial task. Since the 'classical' GA was first proposed by Holland in 1975 as an efficient, easy to use tool which can be applicable to a wide range of problems (Holland, 1975), many variant forms of GAs have been suggested often tailored to specific problems (Michalewicz, 1996). However, it is not always easy for the user to select the appropriate GAs parameters for a particular problem at hand because of the huge number of choices available. At present, there is a little theoretical guidance on how to select the suitable GAs parameters for a particular problem (Michalewicz, 1996). Still another problem is that the natural selection strategy used by GAs is not immune from failure. To cope with the above limitations, an extremely versatile and effective function optimizer called Breeder Genetic Algorithm (BGA) was recently proposed (Muhlenbein, 1994). BGA is inspired by the science of breeding animals. The main idea is to use a selection strategy based on the concept of animal breeding instead of "natural selection" (Irhamah & Ismail, 2009). The assumption behind this strategy is as follows: "mating two individuals with high fitness is more likely to produces an offspring of high fitness than mating two randomly selected individuals"

Some of the features of BGA are:

- BGA uses real-valued representation as opposed to binary representation used in classical GAs. • BGA only requires a few parameters to be chosen by the user.
- The selection technique used is (always) truncation, whereby a selected top T% of the fittest individuals are chosen from the current generation and goes through recombination and mutation to form the next generation. The rest of the individuals are discarded.

The main advantage of using BGA is its simplicity with regard to the selection method (Irhamah & Ismail, 2009) and the fewer parameters to be chosen by the user. However, there is a price to pay for this simplicity. Since only the best individuals are selected in each generation to produce the children for the next generation, there is a likelihood of premature convergence. As a result, BGA may converge to local optimum rather than the desired global one. It should be mentioned that most of the Evolutionary Algorithms including GA have problems with premature convergence to a certain degree. The general way to deal with this problem is to apply mutation to a few randomly selected individuals in the population. In this work, instead of a fixed mutation rate, we have used adaptive mutation strategy (Green, 2005), (Sheetekela & Folly, 2010). This means that the mutation rate is not fixed but varies according to the convergence and performance of the population. In general, even with fixed mutation rate, BGA may still perform better than GA as discussed in (Irhamah & Ismail, 2009). The application of Evolutionary Algorithm to design power system stabilizer for damping low frequency oscillations in power systems has received increasing attention in recent years, see for example, (Wang, et al 2008), (Chuang, & Wu, 2006), (Chuang, & Wu, 2007), (Eslami, et al 2010), (Hongesombut, et al 2005), (Folly, 2006), and (Hemmati, et al 2010). Low frequency oscillations in power systems arise due to several causes. One of these is the heavy transfer of power over long distance. In the last few years, the problems of low frequency oscillations are becoming more and more important. Some of the reasons for this are: a. Modern power systems are required to operate close to their stability margins. A small disturbance can easily reduce the damping of the system and drive the system to instability. b. The deregulation and open access of the power industry has led to more power transfer across different regions. This has the effect of reducing the stability margins. For several years, traditional control methods such as phase compensation technique (Hemmati et al, 2010), root locus (Kundur, 1994), pole placement technique (Shahgholian & Faiz, 2010), etc. have been used to design Conventional PSSs (CPSSs). These (CPSSs) are widely accepted in the industry because of their simplicity. However, conventional controllers cannot provide adequate damping to the system over a wide range of operating conditions. To cover a wide range of operating conditions when designing the PSSs several authors have proposed to use multi-power conditions, whereby the PSS parameters are optimized over a set of specified operating conditions using various optimization techniques such as sensitivity technique (Tiako & Folly, 2009), (Yoshimura& Uchida, 2000),

Differential Evolutionary (Wang, et al 2008), hybrid Differential Evolutionary (Chuang, & Wu, 2006), (Chuang, & Wu, 2007), Particle Swarm Optimization (Eslami, et al 2010), Population-Based Incremental Learning (Folly,2006), (Sheetekela, 2010), etc. In this chapter, Breeder Genetic Algorithm (BGA) with adaptive mutation is used for the optimization of the parameters of the Power System Stabilizer (PSSs). An eigenvalue based objective function is employed in the design such that the algorithm maximizes the lowest damping ratio over specified operating conditions. A single machine infinite bus system is used to show the effectiveness of the proposed method. For comparison purposes, Genetic Algorithms (GAs) based PSS and the Conventional PSS (CPSS) are included. Frequency and time domain simulations show that BGA-PSS performs better than GA-PSS and CPSS under both small and large disturbances for all operating conditions considered in this work. GA-PSS in turn gives a better performance than the Conventional PSS (CPSS).

BACKGROUND THEORY TO BREEDER GENETIC ALGORITHM

BGA is a relatively new evolution algorithm. It is similar to GAs with the exception that it uses artificial selection and has fewer parameters. Also, BGA uses real-valued representation as opposed to GAs which mainly uses binary and sometimes floating or integer representation. In this work, a modified version of BGA called Adaptive Mutation BGA is used (Green, 2005), (Sheetekela & Folly, 2010). Truncation selection method is adopted whereby a top T% of the fittest individuals are chosen from the current population of N individuals and goes through recombination and mutation to form the next generation. The rest of the individuals are discarded. In truncation method, the fittest individual in the population called an ellist is guaranteed a place in the next generation. The other top (T-1) % goes through recombination and mutation to form up the rest of the individuals in the next generation. The process is repeated until an optimal solution is obtained or the maximum number of iteration is reached.

Recombination

Recombination is similar to crossover in GAs (Michalewicz, 1996). The Breeder Genetic Algorithm proposed in this work allows various possible recombination methods to be used, each of them searching the space with a particular bias. Since there is no prior knowledge as to which bias is likely to suit the task at hand, it is better to include several recombination methods and allow selection to do the elimination. Two recombination methods were used in this work: volume and line recombination (Sheetekela, 2010). In volume recombination, a random vector r of the same length as the parent is generated and the child zi is produced by the following expression.

$$z_i = r_i x_i + (1 - r_i)y_i$$

$$(1)$$

where x_1 and y_1 are the two parents. In other words, the child can be said to be located at a point inside the hyper box defined by the parents as shown in Figure. 1. In line recombination, a single uniformly random number r is generated between 0 and 1, and the child is obtained by the following expression (Green, 2005)

$$z_i = r x_i + (1 - r)y_i$$

$$(2)$$

where x_1 and y_1 are the two parents. In light of this, a child can be said to be located at a randomly chosen point on a line connecting the two parents as shown in Figure.2.

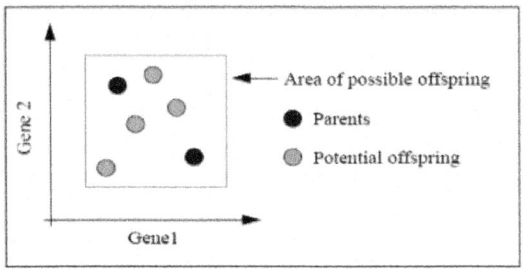

Figure. 1: Volume recombination

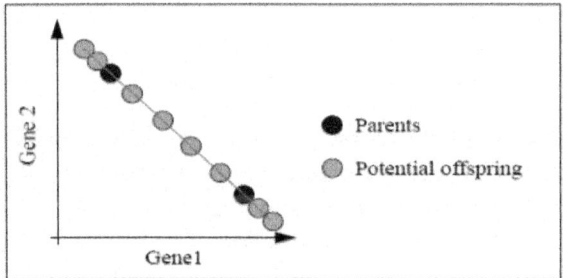

Figure 2: Line recombination

Mutation One problem that has been of concern in GAs is premature convergence, whereby a good but not optimal solution will come to dominate the population. In other words, the search may well converge to local optimum than the desired global one. This problem can be eliminated by adding a small vector of normally-distributed zero-mean random numbers (say with a standard deviation R) to each child before inserting it into the population. The magnitude of the standard deviation R of the vector is very critical, as small R might lead to premature converge and large R might impair the search and reduce its ability to converge optimally. Therefore, it's better to use an adaptive approach whereby the rate of mutation is modified during the course of the search. We set R to the nominal rate R_{nom}. The population is divided into two halves X and Y. A mutation rate of 2Rnom is applied to X whereas a mutation of R_{nom}/2 is applied to Y. The mutation rate R_{nom} is adjusted depending on the population (X or Y) that is producing better and fitter solutions on average. If X individuals are fitter, then the mutation rate R_{nom} is increased slightly by say 10%. If Y is fitter then the mutation rate, R_{nom} is reduced by a similar amount.

Test model

The power system considered is a single machine infinite bus (SMIB) system as shown in Figure. A. 1 of Appendix 8.2.1. The generator is connected to the infinite bus through a doublecircuit transmission line. The generator is modeled using a 6th order machine model, and is equipped with an automatic voltage regulator (AVR) which is represented by a simple exciter of first order differential equation as given in the Appendix 8.1.4. The block diagram of the AVR is shown in Figure. A. 2 of Appendix 8.2.2. A supplementary controller also known as power system stabilizer (PSS) is to be designed to damp the system's oscillations. The block diagram of the PSS is shown in Figure. A.3 of Appendix 8.2.3. The non-linear differential equations of the system are linearized around the nominal operating condition to form a set of linear equations as follows:

$$\begin{cases} \dfrac{d}{dt}x = Ax + Bu \\ y = Cx + Du \end{cases}$$

$$(3)$$

where: A is the system state matrix, B is the system input matrix, C is the system output matrix and D is the feed-forward matrix x is the vector of the system states, u is the vector of the system inputs and y is the vector of the system outputs. In this work $x = [\Delta\delta \ \Delta\omega \ \Delta\psi_{fd} \ \Delta\psi_{1d} \ \Delta\psi_{1q} \ \Delta\psi_{2q} \ \Delta E_{fd}]$; $u = [\Delta T_m \ \Delta V_{ref}]$; $y = \Delta\omega$; the rotor angle deviation, $\Delta\omega$ is the speed deviation, $\Delta\psi fd$ is the field flux linkage deviation, $\Delta\psi ld$ is d-axis amortisseur flux linkage deviation, $\Delta\psi_{1q}$ is the 1st q-axis amortisseur flux linkage deviation, $\Delta\psi_{2q}$ is the 2nd q-axis amortisseur flux linkage deviation, ΔE_{fd} is the exciter output voltage deviation. ΔT_m is the mechanical torque deviation and ΔV_{ref} is the voltage reference deviation.

Several operating conditions were considered for the design of the controllers. These operating conditions were obtained by varying the active power output, P_e and the reactive power Q_e of the generator as well as the line reactance, X_e. However, for simplicity, only three operating conditions will be presented in this paper. These operating conditions are listed in the Table 1 together with the open loop eigenvalues and their respective damping ratios in % in brackets.

Table 1: Selected operating conditions with open-loop eigenvalues

case	Active Power Pe [p.u]	Reacctive Power Qe [p.u]	Line reactance Xe [p.u]	Eigenvalues (Damping ratio)
1	1.1000	0.4070	0.7000	-0.2894 ± j5.2785 (0.0547)
2	0.5000	0.1839	1.1000	-0.3472 ± j4.3271 (0.0800)
3	0.9000	0.3372	0.9000	-0.2704 + j4.7212 (0.0572)

FITNESS FUNCTION

The fitness function is used to provide the measure of how individuals performed. In this instance, the problem domain was that the PSS parameters should stabilize the system simultaneously over a certain range of specified operating conditions. The PSS which parameters are to be optimized has a structure similar to the conventional PSS (CPSS) as shown in Figure. A. $_3$. of Appendix 8.2.3. There are three parameters KS, T_1 and T_2 that are to be optimized, where Ks is the PSS gain and T_1 and T_2 are lead-lag time constants. T_w is the washout time constant which is not critical and therefore has not been optimized. The fitness function that was used is to maximize the lowest damping ratio. Mathematically the objective function is formulated as follows:

$$val = \max(\min(\varsigma_{ij}))$$

(4)

where i = 1,2, ... n , j =1, 2, ...m

$$\varsigma_{ij} = \frac{-\sigma_{ij}}{\sqrt{\sigma_{ij}^2 + \omega_{ij}^2}}$$

ς_{ij} is the damping ratio of the ith eigenvalue of the jth operating conditions. The number of the eigenvalues is n, and m is the number of operating conditions. σ_{ij} and ω_{ij} are the real part and the imaginary part (frequency) of the eigenvalue, respectively.

PSS design

The following parameter domain constraints were considered when designing the PSS.

$$0 < K_s \le 20$$

$$0.001 \le T_i \le 5$$

where K_s and T_i denote the controller gain and the lead lag time constants, respectively .

BGA-PSS

The following BGA parameters have been used during the design

- Population: 100
- Generation: 100
- Selection: Truncation selection (i.e., selected the best 15% of the population)
- Recombination: Line and volume
- Mutation initial R_{nom}: 0.01

The parameters of the BGA-PSS are given in Table A.1 of Appendix 8.2.3.

GA-P15 Folly_secondSS

The following GA parameters have been used during the design

- Population: 100
- Generation: 100
- Selection: Normalized geometric
- Crossover: Arithmetic
- Mutation: Non-uniform

More information on the selection, crossover and mutation can be found in (Michalewicz, 1996), (Sheetekela & Folly, 2010). The parameters of the GA-PSS are given in Table A.1 of Appendix 8.2.3.

Conventional-PSS

The Conventional PSS (CPSS) was designed at the nominal operating condition using the phase compensation method. The phase lag of the system was first obtained, which was found to be 20o, thus only a single lead-lag block was used for the PSS. After obtaining the phase lag, a PSS with a phase lead was designed using the phase compensation technique. The final phase lead obtained was approximately 18°, thus giving the system a slight phase lag of 2°. Once the phase lag is improved, then the damping needed to be improved as well by varying the gain K_S. The parameters of the CPSS are given in Table A.1 of Appendix 8.2.3.

SIMULATION RESULTS

Eigenvalue analysis

Under the assumption of small-signal disturbance (i.e, small change in Vref or Tm), the eigenvalues of the system are obtained and the stability of the system investigated. Table 2 shows the eigenvalues of the system for the different PSSs. The damping ratios are shown in brackets. For all of the cases, it can be seen that on average, BGA-PSS provides more damping to the system than GA-PSS. On the other hand, GA-PSS performs better than CPSS. For example for case 1, BGA-PSS provides a damping ratio of 50% as compared to 48.85% for GA-PSS and 44.93% for CPSS. This means that, BGA gives the best performance. Likewise, BGA provides better damping ratios for cases 2 and 3.

Table 2. Closed-loop eigenvalues

case	BGA-PSS	GA-PSS	CPSS
1	-3.0664 ±j 5.3117 (0.5000)	-2.9208 ± j5.2172 (0.4885)	-1.9876 ± j3.9516 (0.4493)
2	-1.2793 ± j4.3024 (0.2850)	-1.2305± j4.2616 (0.2774)	-0.9529 ± j3.9443 (0.2348)
3	-2.1245 + j4.6503 (0.4155)	-2.0268 + j4.5784 (0.4048)	-1.3865 + j3.8881 (0.3359)

It should be mentioned that a maximum damping ratio of 50% was imposed on the BGA and GA, otherwise, their damping ratios could have been higher. If the damping of the electromechanical mode is too high this could negatively affect other modes in the system.

Large disturbance

A large disturbance was considered by applying a three-phase fault to the system at 0.1 seconds. The fault was applied at the sending-end of the system (near bus 1 on line 2) for 200ms. The fault was cleared by disconnecting line 2. Figure. 3 to Figure. 5 show the speed responses of the system. Figureure 3 shows the speed responses of the generator for case 1. When the system is equipped with GA-PSS and BGA-PSS it settles around 3 seconds. On the other hand, the settling time of the system equipped with the CPSS is more than doubled (6 seconds). In addition, the subsequent oscillations are larger than those of BGA and GA PSSs. Figureure 4 shows the speed responses for case 2. The system equipped with CPSS is seen to have bigger oscillations as compared to the system equipped with BGA-PSS and GA-PSS. With both BGA and GA PSSs, the system settled in approximately 3.5 sec., whereas CPSS takes more than 6 sec. to settle down. The performances of the BGA-PSS and GA-PSS are quite similar, even though the BGA-PSS performs slightly better than the GA- PSS. Figureure 5 shows the speed responses of the system for case 3. It can be seen that the system equipped with BGA and GA PSS settled in less than 4 sec compared to more than 6 sec. for the CPSS. With CPSS, the system has large overshoots and undershoots.

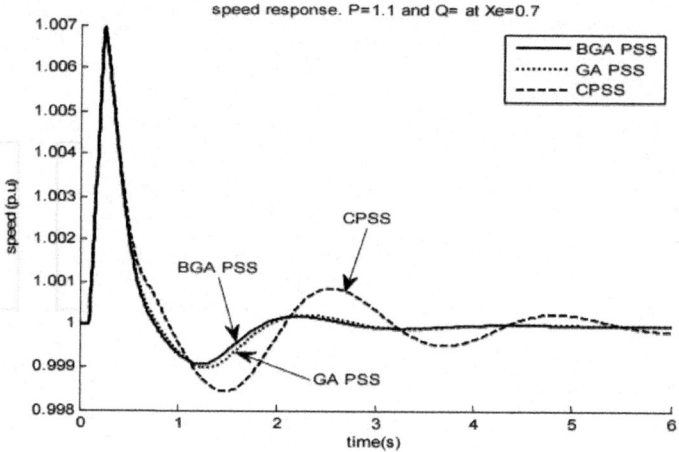

Figure. 3: Speed response of case 1 under three-phase fault

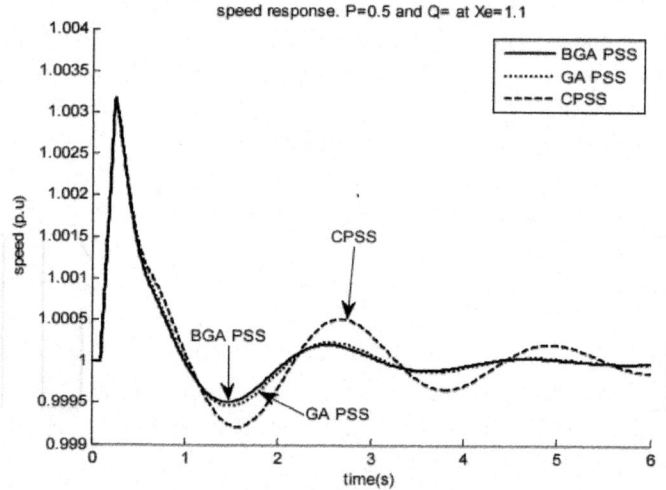

Figure. 4: Speed responses of case 2 under three-phase fault

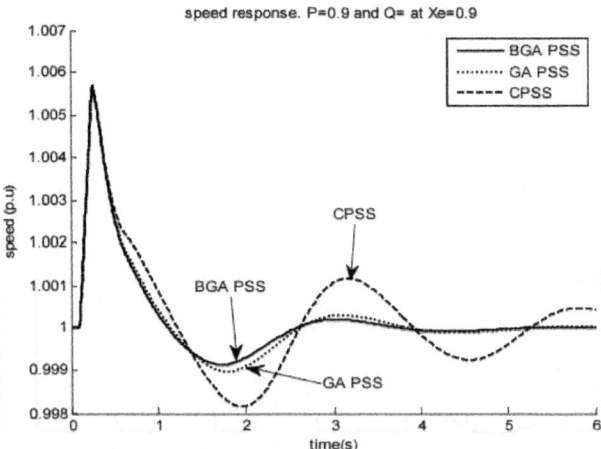

Figure 5: Speed responses of case 3 under three-phase fault

CONCLUSION

Breeder Genetic Algorithms is an extremely versatile and effective function optimizer. The main advantage of BGA over GA is the simplicity of the selection method and the fewer genetic parameters. In this work, adaptive mutation has been used to deal with the problem of premature convergence in BGA. The effectiveness of the proposed approach was demonstrated by the time and frequency domain simulation results. Eigenvalue analysis shows that the BGA based controller provides a better damping to the system for all operating conditions considered than a GA based controller. The conventional controller provides the least damping to all the operating conditions considered. The robustness of the BGA controller under large disturbance was also investigated by applying a three-phase fault to the system. Further research will be carried out in the direction of using multiobjective functions in the optimization and using a more complex power system model.

APPENDIX

Generator and Automatic Voltage Regulator (AVR) equations

Swing equations

$$\frac{d}{dt}\Delta\omega = \frac{1}{2H}(T_m - T_e - K_D\Delta\omega)$$

$$\frac{d}{dt}\Delta\delta = \omega_0\Delta\omega$$

where

δ is the rotor angle in rad

ω is the synchronous speed in per-unit (p.u.)

ω_0 is the synchronous speed in rad/sec

H is the inertia constant in sec.

T_m is the mechanical torque in p.u.

T_e is the mechanical torque in p.u.

K_D is the damping coefficient in torque/ p.u.

Rotor circuit equations

$$\frac{d}{dt}\psi_{fd} = \omega_0\left(E_{fd} - \frac{R_{fd}}{L_{fd}}i_{fd}\right)$$

$$\frac{d}{dt}\psi_{1d} = -\omega_0 R_{1d}i_{1d}$$

$$\frac{d}{dt}\psi_{1q} = -\omega_0 R_{1q}i_{1q}$$

$$\frac{d}{dt}\psi_{2q} = -\omega_0 R_{2q}i_{2q}$$

Where

ψ_{fd}, ψ_{1d}, ψ_{1q}, ψ_{2q}, E_{fd} are the same as defined in section 3.

R_{fd}, L_{fd}, are the field winding resistance and inductance, respectively.

R_{1d}, is the d-axix amortisseur resistance.

R_{1q}, is the 1st q-axix amortisseur resistance.

R_{2q} is the 2_{nd} q-axix amortisseur resistance.

The rotor currents are expressed a follows:

$$i_{fd} = \frac{1}{L_{fd}}(\psi_{fd} - \psi_{ad})$$

$$i_{1d} = \frac{1}{L_{1d}}(\psi_{1d} - \psi_{ad})$$

$$i_{1q} = \frac{1}{L_{1q}}(\psi_{1q} - \psi_{aq})$$

$$i_{2q} = \frac{1}{L_{2q}}(\psi_{2q} - \psi_{aq})$$

where

ψ_{fd}, ψ_{1d}, ψ_{1q}, ψ_{2q} are defined as before

ψ_{ad}, ψ_{aq}, are the mutual flux linkages in the d and q axis, respectively.

L_{1d} is the d-axix amortisseur inductance.

L_{1q} is the 1st q-axix amortisseur inductance.

L_{2q} is the 2nd q-axix amortisseur inductance.

Electrical torque

The electrical torque is expressed by the following:

$$T_e = \psi_d i_q - \psi_q i_d$$

where ψ_d, and ψ_q are the d and q axis flux linkages, respectively.

AVR equations

$$\frac{d}{dt}E_{fd} = \frac{K_A}{T_A}(V_{ref} - V_t) - \frac{E_{fd}}{T_A}$$

where K_A and T_A are the gain and time constant of the A_{VR}. V_t is the terminal voltage of the generator. In this work K_A=200 and T_A = 0.05 sec.

Power system model, AVR parameters and PSS block diagram and parameters

Power system model diagram

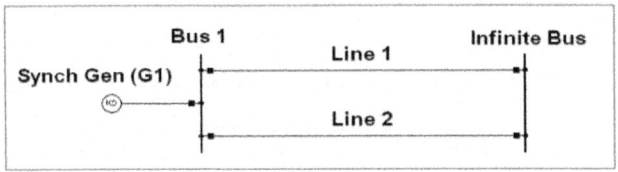

Figure. A1: System model- Single-Machine Infinite Bus (SMIB)

Block diagram of the Automatic Voltage Regulator (AVR)

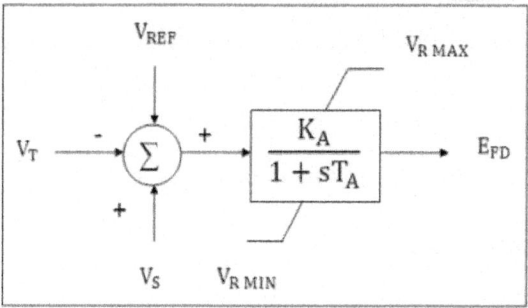

Figure. A2: Automatic voltage regulatore structure

Block diagram and parameters of the PSSs

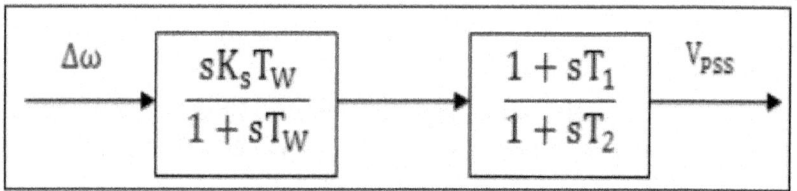

Figure. A3: Power system stabilizer structure

In Figure. A3, V_{PSS} is the output signal of the PSS, while $\Delta\omega(s)$ is the input signal, which in this case is the speed deviation.

Table A1: PSS parameters

PSSs	K_s	T_1	T_2	T_w
CPSS	9.7928	1.1686	0.2846	2.5000
GA-PSS	13.7358	3.5811	1.2654	2.5000
BGA-PSS	18.8838	3.7604	1.7390	2.5000

Generator's parameters

X_l =0.0742 p.u, , X_d=1.72 p.u,, X'_d=0.45 p.u,, X''_d=0.33 p.u, T'_{d0}=6.3sec., T''_{d0} = 0.033 p.u,,
X_q =1.68 p.u,, X'_q =0.59 p.u,, X''_q =0.33 p.u, T'_{q0} =0.43 sec

T''_{q0} = 0.033sec., H = 4.0sec

Pseudo code for BGA generator's parameters

```
Begin
    Randomly initialize a population of N individuals;
    Initialize mutation rate R_nom
    While termination criterion not met
        evaluate goodness of each individuals
        save the best individual in the new population
        select the best T% individuals and discarding the rest;
        for I =1 to N-1 do
            randomly select two individuals among the T% best individual
            recombine the two parents to obtain one offspring
        end
        divide the new population into two halves (X and Y)
        apply mutation rate r_nom/2 to X and 2 R_nom to Y
        evaluate the average fitness value for the two half population (X and Y)
        If X performs better than Y; assign r= R_nom -0.1 r_nom;
        If Y performs better than X; assign r= R_nom + 0.1 r_nom;
    end
end
```

ACKNOWLEDGMENT

The authors would like to acknowledge the financial support of THRIP and TESP.

REFERENCES

1.   Baluja, S. (1994). Population-Based Incremental Learning: A method for Integrating Genetic Search Based Function Optimization and Competitive Learning. Technical Report CMU-CS-94-163,

2.   Carnegie Mellon University. Chuang, Y.S . & Wu, C. J. (2006). A Damping Constant Limitation Design of Power System Stabilizer using Hybrid Differential Evolution. Journal of Marine Science and Technology. Vol..14, No. 2, pp. 84-92. Chuang, Y.S . & Wu, C. J. (2007).

3.   Novel Decentralized Pole Placement Design of Power System Stabilizers using Hybrid Differential Evolution. International Journal of Mathematics and Computer in Simulation. Vol..1, No. 4, pp. 410-418.

4.   Eslami, M., Shareef, H., Mohamed, A., & Ghohal, S. P. (2010). Tuning of Power System Stabilizer using Particle Swarm Optimization with Passive Congregation. International Journal of Physical Sciences. Vol..5, No. 17, pp. 2574-2589. Goldberg, D. E. (1989).

5.   Genetic Algorithms in Search, Optimization & Machine Learning. Addison-Wesley; 1989.

6.   Green, J. (2005). The Idea Behind Breeder Genetic Algorithm, Department of Electrical Eng., University of Cape Town Folly KA (2006)

7.   Design of Power System Stabilizer: A Comparison Between Genetic Algorithms and Population Based Incremental Learning, 2006 IEEE Power Engineering Society General Meeting, ISSN: 19325517.

8.   Hemmati, R. , Mojtaba, S, Boroujeni, S & Abdollahi, M. (2010). Comparison of Robust and Intelligent Based Power System Stabilizers. International Journal of the Pysical Science. Vol. 5, No. 17, pp. 2564-2573.

9.   Holland, J. H. (1975). Adaptation in Nature and Artificial Systems, University of Michigan Press, Ann Arbor. Hongesombut, K., Dechanupaprittha S., Mitani, Y. & Ngamroo, I (2005).

10. Robust Power System Stabilizer Tuning based on Multiobjective Design Using Hirerarchical & Parallel Micro Genetic Algorithm. Proc. of the 15th Power Syst. Comp. Conf., Liege. Irhamah & Ismail, Z. (2009)

11. A Breeder Genetic Algorithm for Vehicle Routing Probelm with Stochastic Demands. Journal of Applied Research, Vol.5, No.11, pp. 1998-2005

12. Kundur, P. (1994). Power System Stability and Control Mc Graw Hill; 1994. Michalewicz, Z. (1996) Genetic Algorithms+Data Structure =Evolution Programs. 3rd Ed. Springer-Velag Mitchell, M. (1996), An introduction to genetic algorithms, The MIT Press. Muhlenbein, H. (1994),

13. The Breeder Genetic Algorithm-A Provable Optimal Search Algorithm and its Application, Available from http://ieeexplore.ieee.org Sheetekela, S & Folly KA (2010).

14. Breeder Genetic Algorithm for Power System Stabilizer Design. 2010 IEEE Congress on Evolutionary Computation (CEC), Barcelona, Spain. Sheetekela (2010).

15. Design of Power System Stabilizer using Evolutionary Algorithms, MSc. Thesis Department of Electrical Eng., University of Cape Town Shahgholian, G. & Faiz, J. (2010).

16. The effect of Power System Stabilizer on Small Signal Stability in Single Machine Infinite-Bus. Internatonal Journal of Electrical and Power Engineering.Vol.14, No. 2, pp. 45-53.

17. Sundareswara, K. & Begum, S. R. (2004). Genetic Tuning of a Power System Stabilizer. Euro. Trans. Electr. Vol..14, pp. 151-160. Tiako, R. & Folly, K A. (2009).

18. Investigation of Power System Stabilizer Parameters Optimisation using multi-power flow conditions. Australian Journal of Electrical & Electronics Engineering. Vol..5, No. 3, pp. 237-244.

19. Yoshimura, K. & Uchida, N. (2000). Optimization of P+w Parameters for Stability and Robustness Enhancement in a Multimachine Power System. Electrical Engineeringin Japan. Vol..131, No. 1, pp. 19-31

20. (Translated from Denki Gakkai Ronbunshi, Vol. 11, Nov. 1198, pp. 1312-1320). Wang, Z. Chung, C.Y., & Wong, C.T. (2008). Robust Power System Stabilizer Design under multi-operating conditions using Differential Evolution. IET Generation, Transmission & Distribution. Vol..2, No. 5, pp. 690-700.

Chapter 10

PERFORMANCE STUDY OF CULTURAL ALGORITHMS BASED ON GENETIC ALGORITHM WITH SINGLE AND MULTI POPULATION FOR THE MKP

Deam James Azevedo da Silva[1], Otávio Noura Teixeira[2] and Roberto Célio Limão de Oliveira[1]

[1]Universidade Federal do Pará (UFPA), Brazil

[2]Centro Universitário do Estado do Pará (CESUPA) Brazil

INTRODUCTION

Evolutionary Computation (EC) is inspired from by evolution that explores the solution space by gene inheritance, mutation, and selection of the fittest candidate solutions. Since their inception in the 1960s, Evolutionary Computation has been used in various hard and complex optimization problems in search and optimization such as: combinatorial optimization, functions optimization with and without constraints, engineering problems and others (Adeyemo, 2011). This success is in part due to the unbiased nature of their operations, which can still perform well in situations with little or no domain knowledge (Reynolds, 1999). The basic EC framework consists of fairly simple steps like definition of encoding scheme, population generation method, objective function, selection strategy, crossover and mutation (Ahmed & Younas, 2011). In addition, the same procedures utilized by EC can be applied to diverse problems with relatively little reprogramming. Cultural Algorithms (CAs), as well as Genetic Algorithm (GA), are evolutionary models that are frequently employed in optimization problems. Cultural Algorithms (CAs) are based on knowledge of an evolutionary system and were introduced by Reynolds as a means of simulating cultural evolution (Reynolds, 1994). CAs algorithms implements a dual mechanism of inheritance where are inherited characteristics of both the level of the population as well as the level of the area of belief space (culture). Algorithms that use social learning are higher than those using individual learning, because they present a better and faster convergence in the search for solutions (Reynolds, 1994). In CAs the characteristics and behaviors of individuals are represented in the Population Space. This representation can support any population-based computational model such as Genetic Algorithms, Evolutionary Programming, Genetic Programming, Differential Evolution, Immune Systems, among others (Jin & Reynolds, 1999). Multidimensional Knapsack Problem (MKP) is a well-known nondeterministic-polynomial time-hard combinatorial optimization problem, with a wide range of applications, such as cargo loading, cutting stock problems, resource allocation in computer systems, and economics (Tavares et al., 2008). MKP has received wide attention from the operations research community, because it embraces many practical problems. In addition, the MKP can be seen as a general model for any kind of binary problems with positive coefficients (Glover & Kochenberger, 1996). Many researchers have proposed the high potential of the hybrid-model for the solution of problems (Gallardo et al., 2007). The algorithms presented in this work to solve MKP are a combination of CAs with a Multi Population model. The Multi Population model is the division of a population into several smaller ones, usually called the island model. Each sub-population runs a standard sequential evolution proceeds, as if it were isolated from the rest, with occasional migration of individuals between sub-populations (Tomassini, 2005). In order to conduct an investigation to discover improvements for MKP, this work is centered in the knowledge produced from CAs through the evolutionary process that utilizes a population-based Genetic Algorithm model, using various MKP benchmarks found in the literature. In addition, there is an interest in investigating how to deal with the Cultural Algorithms considering a population-based in Genetic Algorithms. So as to compare test results, we implemented the follows algorithms: the standard cultural algorithm with Single Population (also known as standard CA or CA-S) and Cultural Algorithm with Multi Population defined as CA-IM with two versions: CA-IM_1 which has fixed values for genetic operators (recombination and mutation) and CA-IM_2 which does not have fixed values for genetic operators because these values are generated randomly. In order to evaluate the performance of the CA-IM algorithms, some comparison testing will be conducted with other two algorithms based on Distributed GA, called DGA and DGASRM (Aguirre et al., 2000). The outline of the paper is as follows: in Section 2, a description with formal definition of the MKP problem and an overview of Cultural Algorithms are presented. Section 3 shows an alternative approach that explores the multi population model with Cultural Algorithms and explores how the interaction process occurs among various sub-populations. Our experimental results are shown in Section 4 and finally we show some conclusions in Section 5.

BACKGROUND

Since the introduction of the Knapsack problems some algorithm techniques such as brute force, conventional algorithms, dynamic programming, greedy approach and approximation algorithm have been proposed (Ahmed & Younas, 2011). Evolutionary algorithms (EAs) have been widely applied to the MKP and have shown to be effective for searching and finding good quality solutions (Chu & Beasley, 1998). It is important to note that MKP is considered a NP hard problem; hence any dynamic programming solution will produce results in exponential time. In the last few years, Genetic Algorithms (GAs) have been used to solve the NP-complete problems and have shown to be very well suited for solving larger Knapsack Problems (Fukunaga & Tazoe, 2009; Gunther, 1998; Sivaraj & Ravichandran, 2011). For larger knapsack problems, the efficiency of approximation algorithms is limited in both solution quality and computational cost (Ahmed & Younas, 2011). Spillman's experiment, which applies the GA to the knapsack problem, shows that the GA does not have a good performance in relatively small size problem, but works quite well in problems that include a huge number of elements (Spillman, 1995). There are many packing problems where evolutionary methods have been applied. The simplest optimization problem and one of the most studied is the onedimensional (zero–one or 0-1) knapsack problem (Ahmed & Younas, 2011), which given a knapsack of a certain capacity, and a set of items, each one having a particular size and value, finds the set of items with maximum value which can be accommodated in the knapsack. Various real-world problems are of this type: for example, the allocation of communication channels to customers who are charged at different rates (Back et al., 1997). During a study of 0-1 knapsack, a number of extensions and variants have been developed such as (Ahmed & Younas, 2011): Multiple Knapsack Problems (MKP), Multidimensional Knapsack Problems (MDKP), Multi Choice Knapsack Problems (MCKP) and Multiple Choice Multidimensional Knapsack Problems (MMKP). It is also important to consider other extensions such as (Chu & Beasley, 1998): Multiconstraint Knapsack Problem, and also the term "Multidimensional Zero-one Knapsack Problem". Using alternative names for the same problem is potentially confusing, but since, historically, the designation MKP has been the most widely used (Chu & Beasley, 1998). Consequently, Multidimensional Knapsack Problem (MKP) is the designation selected for this work. In our previous research it was introduced a Multi Population Model on the cultural structure identified as "Multi Population Cultural Genetic Algorithm" (MCGA) (Silva & Oliveira, 2009). In MCGA model several sub-populations are connected with as ring structure, where the migration of individuals occurs after a generation interval (according to the migration based on parameter interval) with best-worst migration policy implementation. The results were satisfactory in relation to other algorithms in the literature. In another research two versions of Distributed GA (DGA) are presented as follows: standard Distributed GA (DGA) and an improved DGA (DGA-SRM), which two genetic operators are applied in parallel mode to create offspring. The term SRM represents "Self-Reproduction with Mutation", that is applied to various 0/1 multiple knapsack problems so as to improve the search performance (Aguire et al., 2000). Hybridization of memetic algorithms with Branch-and-Bound techniques (BnB) is also utilized for solving combinatorial optimization problems (Gallardo et al., 2007). BnB techniques use an implicit enumeration scheme for exploring the search space in an "intelligent" way. Yet another research utilizes adaptive GA for 0/1 Knapsack problems where special consideration is given to the penalty function where constant and self-adaptive penalty functions are adopted (Zoheir, 2002). Fitness landscape analysis techniques are used to better understand the properties of different representations that are commonly adopted when evolutionary algorithms are applied to MKP (Tavares et al., 2008). Other investigation utilizes multiple representations in a GA for the MKP (Representation-Switching GA) know as RSGA (Fukunaga, 2009). Other recent works consider two heuristics and utilize them for making comparisons to the well-known multiobjective evolutionary algorithms (MOEAs) (Kumar & Singh, 2010). While comparing MOEAs with the two heuristics, it was observed that the solutions obtained by the heuristics are far superior for larger problem instances than those obtained by MOEAs

Multidimensional Knapsack Problem

As mentioned earlier, the MKP is a well-known nondeterministic-polynomial time-hard combinatorial optimization problem, with a wide range of applications (Tavares et al., 2008). The classical 0-1 knapsack problem is one of the most studied optimization and involves the selection of a subset of available items having maximum profit so that the total weight of the chosen subset does not exceed the knapsack capacity. The problem can be described as follows: given two sets of n items and m knapsacks constraints (or resources), for each item j, a profit p_j is assigned, and for each constraint i, a consumption value r_{ij} is designated. The goal is to determine a set of items that maximizes the total profit, not exceeding the given constraint capacities c_i. Formally, this is stated as follows (Tavares et al., 2008):

$$\text{Maximize} \quad \sum_{j=1}^{n} p_j x_j , \tag{1}$$

$$\sum_{j=1}^{n} r_{i,j} x_j \le c_{i,} \quad i = 1,\dots,m$$

Subject (2)

$$x_j \in \{0,1\}, \quad j = 1,\dots,n$$
(3)

With

$$p_j > 0, \ r_{i,j} \ge 0 \text{ and } c_i \ge 0$$
(4)

The knapsack constraint is represented by each of the m constraints described in E_q. (2). The decision variable is the binary vector $x = (x_1,\dots,x_n)$. Each item j is mapped to a bit and when $x_j = 1$, the corresponding item is considered to be part of the solution. The special case of m =1 is generally known as the Knapsack Problem or the Unidimensional Knapsack Problem

For single constraint the problem is not strongly NP-hard and effective approximation algorithms have been developed for obtaining near-optimal solutions. A review of the single knapsack problem and heuristic algorithms is given by Martello and Toth (Martello & Toth, 1990). Exact techniques and exhaustive search algorithms, such as branch-andbound, are only of practical use in solving MKP instances of small size since they are, in general, too time-consuming (e.g., instances with 100 items or less, and depending on the constraints).

Evolutionary approach for the MKP

In a resolution of specific problems that implements an Evolutionary Algorithm, as for example, a simple Genetic Algorithm (GA), it is necessary the definition of five components (Agapie et al., 1997). The first component is the genotype or a genetic representation of the potential problem (individual representation scheme). The second is a method for creating an initial population of solutions. The third is a function verifying the fitness of the solution (objective function or fitness function). The fourth are genetic operators and the fifth are some constant values for parameters that are used by the algorithm (such as population size, probability of applying an operator, etc.).

Genotype

The natural representation of the MKP would be the binary representation, in which every bit represents the existence or not of a certain element in the Knapsack. A bit set to 1 indicates that the corresponding item is packed into the knapsack and a bit set to 0 indicates that it is not packed. Hence a typical population of two individuals for a six elements in Knapsack would be represented as showed in Figure 1. Thus, each element has an identification that is given by the bit index. In Figure 1 (a) there are three elements in the knapsack, corresponding to the following positions: 1, 4 and 6. In Figure 1 (b) there are four elements in the knapsack, whose positions are: 2, 3, 5 and 6.

a)   b)

Figure 1: Knapsack example for two chromosomes.

Initial population

The population is the solution representation that consists of a set of codified chromosomes. There are many ways to generate the initial population such as random chromosome or chromosome with the solution closer to the optimum. In most applications the initial population is generated at random.

Evaluation function

In GA each individual is evaluated by fitness function. Some individuals produce more children than others due to their fitness. By this mechanism, individuals that have chromosomes with better fitness have better chances of leaving their genes. This leads to better average performance of the whole population as generations proceed (Ku & Lee, 2001). A feasible vector solution x needs to satisfy constraint (2), otherwise it is infeasible. Hence, a penalty is applied to all infeasible solutions in order to decrease their corresponding "fitness". Therefore, the two types of evaluation functions used in this research are based on static (constant) and adaptive penalty functions. The standard evaluation function for each individual is given by the following expressions:

$$\text{Evaluation }(x) = \sum_{i=1}^{i=n}(x[i]\times p[i]) - Pen(x)$$

(5)

$$\text{Maximum Profit Possible (MaxP)} = \sum_{i=1}^{i=n}p[i]$$

(6)

A vector solution x is optimal when Evaluation (x) $=\text{Max}_p$.

Genetic operators

To implement the GA process, many factors should be considered such as the representation scheme of chromosomes, the mating strategy, the size of population, and the design of the genetic operators such as selection, mutation and recombination (Ku & Lee, 2001)

- Selection - is an operator that prevents low fitness individuals from reproduction and permits high fitness individuals to offspring more children to improve average fitness of population over generations. There are various selections types, such as stochastic remainder, elitism, crowding factor model, tournament, and roulette wheel.

- Recombination or Crossover - is an operator that mixes the chromosomes of two individuals. Typically two children are generated by applying this operator, which are similar to the parents but not same. Crossover causes a structured, yet randomized exchange of genetic material between solutions, with the possibility that the "fittest" solutions generate "better" ones. A crossover operator should preserve as much as possible from the parents while creating an offspring.

- Mutation - introduces totally new individuals to population. It helps extend the domain of search and will restrain the diversity of the population. Mutation involves the modification of each bit of an individual with some probability Pm. Although the mutation operator has the effect of destroying the structure of a potential solution, chances are it will yield a better solution. Mutation in GAs restores lost or unexplored genetic material into the population to prevent the premature convergence of the GA.

The tournament is the selection type chosen for this work since it is more used and it presents good performance. For a binary representation, classical crossover and mutation operators can be used, such as n-point crossover or uniform crossover, and bit-flip mutation. In CAs the influence of information from Belief Space on recombination and mutation process such as: best chromosome or set of best chromosomes information is expected.

Constant values parameters

An Evolutionary Algorithm involves different strategy parameters, e.g.: mutation rate, crossover rate, selective pressure (e.g., tournament size) and population size. Good parameter values lead to good performance. There are three major types of parameter control (Eiben & Smith, 2008):

- deterministic: a rule modifies strategy parameter without feedback from the search (based on some type of a counter);

- adaptive: a feedback rule based on some measure monitoring search progress (quality);

- self-adaptative: parameter values evolve along with the solutions; encoded onto chromosomes they undergo variation and selection.

The implementation of a deterministic parameter control is easier, provided that the parameter values used are tested to verify the best performance.

Cultural algorithms

Cultural Algorithms (CAs) have been developed so as to model the evolution of the cultural component of an evolutionary computational system over time as it accumulates experience

(Reynolds & Chung, 1996). As a result, CAs can provide an explicit mechanism for global knowledge and a useful framework within which to model self-adaptation in an EC system. The CAs are based on knowledge of an evolutionary system that implements a dual mechanism of inheritance. This mechanism allows the CAs to explore as much microevolution as macroevolution. Microevolution is the evolution that happens in the population level. Macroevolution occurs on the culture itself, i.e. the belief space evolution. The belief space is the place where the information on the solution of the problem is refined and stored. It is acquired through the population space over the evolutionary process. The belief space has the goal to guide individuals in search of better regions. In the CAs evolution occurs more quickly than in population without the mechanism of macroevolution. The characteristics and behaviors of individuals are represented in the Population Space and as mentioned earlier the population space can support any population-based computational model such as Genetic Algorithms among others (Jin & Reynolds, 1999). The communications protocols dictate the rules about individuals that can contribute to knowledge in the Belief Space (function of acceptance) and how the Belief Space will influence new individuals (Function of Influence), as shown in Figure 2.

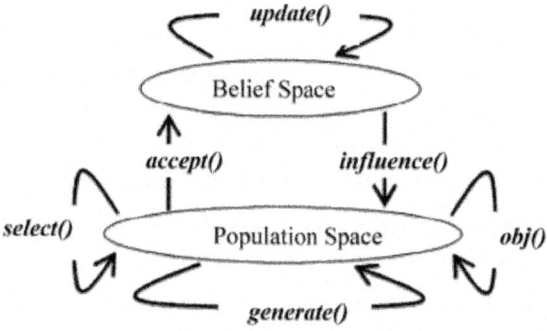

Figure 2: Framework of Cultural Algorithm (Reynolds & Peng, 2004)

The two most used ways to represent knowledge in the belief space are (Reynolds & Peng, 2004): Situational Knowledge and Normative Knowledge. Situational Knowledge represents the best individuals found at a certain time of evolution and it contains a number of individuals considered as a set of exemplars to the rest of the population. The number of exemplars may vary according to the implementation, but it is usually small. For example, the structure used to represent this type of knowledge is shown in Figure 3. Each individual is stored within its parameters and fitness value (Iacoban et al. 2003).

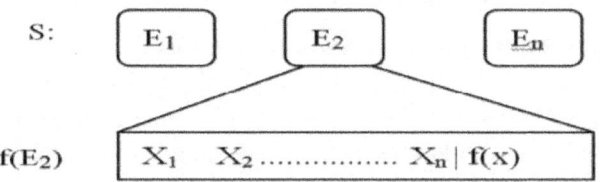

Figure 3: Representation of Situational Knowledge.

The Situational Knowledge is updated when the best individual of the population is found. This occurs when its fitness value exceeds the fitness value of the worst individual stored.

Normative Knowledge represents a set of intervals that characterize the range of values given by the features that make the best solutions (Iacoban et al., 2003). Figure 4 shows the structure used by Reynolds and his students, where are stored the minimum and maximum values on the individual's characteristics.

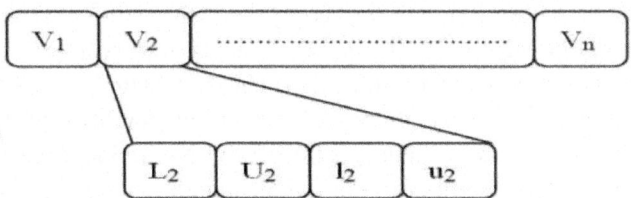

Figure 4: Representation of Normative knowledge

These intervals are used to guide the adjustments (mutations) that occur in individuals. With these minimum values, (li) and maximum (ui), the fitness values are also stored. This value results from the individuals that produced each extreme Li and Ui respectively. The adjustment of the range of Normative Knowledge varies according to the best individual. That is, if the individual was accepted by the acceptance function and its range is less than the range stored in the belief space, the range is adjusted, and vice versa. The resolution of problems produces experiences from individuals in the population space, which are selected to contribute to the acceptance by the belief space, where the knowledge is generalized and stored. In the initial population, the individuals are evaluated by the fitness function. Then, the information on the performance of the function is used as a basis for the production of generalizations for next generations. The experiences of the individuals selected will be used to make the necessary adjustments on the knowledge of the current belief space.

Parallel Genetic Algorithms

The definition of Parallel Genetic Algorithms (PGAs) is related with execution of various GAs in parallel mode. The main goal of PGAs is to reduce the large execution times that are associated with simple genetic algorithms for finding near-optimal solutions in large search spaces and to find better solutions. The PGAs can be implemented through two approaches (Sivanandam, 2007): standard parallel approach and the decomposition approach. In the first approach, the sequential GA model is implemented on a parallel computer by dividing the task of implementation among the processors. The standard parallel approaches are also known as master-slave GAs. In the decomposition approach, the full population exists in distributed form. Other characteristic in the decomposition approach is that the population is divided into a number of sub-populations called demes. Demes are separated from one another and individuals compete only within a deme. An additional operator called migration is used to move the individuals from one deme to another. If the individuals can migrate to any other deme, the model is called island model or Multiple-population GAs when implemented in parallel or distributed environments (Braun, 1991). Migration can be controlled by various parameters like migration rate, topology, migration scheme like best/worst/random individuals to migrate and the frequency of migrations (Sinvanadam, 2007). Other authors classify Parallel Genetic Algorithm in four main categories (Aguirre & Tanaka, 2006): global master-slave, island, cellular, and hierarchical parallel GAs. In a global master-slave GA there is a single population and the evaluation of fitness is distributed among several processors. The important characteristic in a global master-slave GA is that the entire population is considered by genetic operators as selection, crossover and mutation. An island GA, also known as coarse-grained or distributed GA, consists of several sub-populations evolving separately with occasional migration of individuals between subpopulations. A cellular category also known as "fine-grained GA" consists of one spatially structured population, whose selection and mating are restricted to a small neighborhood. The neighborhoods are allowed to overlap permitting some interaction among individuals. Finally, a hierarchical parallel GA category, combines an island model with either a masterslave or cellular GA. The global master-slave GA does not affect the behavior of the algorithm and can be considered only as a hardware accelerator. However, the other parallel formulations of GAs are very different from canonical GAs, especially, with regard to population structure and selection mechanisms. These modifications change the way the GA works, affecting its dynamics and the trajectory of evolution. For example, the utilization of parameters as sub-population size, migration rate and migration frequency are crucial to the performance of island models. Cellular, island and hierarchical models perform as well as or better than canonical versions and have the potential of being more than just hardware accelerators (Aguirre & Tanaka, 2006). A new taxonomy about PGAs is also presented by Nowostawski and Poli (Nowostawski & Poli, 1999). In recent studies about MKP Silva and Oliveira (Silva & Oliveira, 2009) have shown that good results are reached in the benchmark tests when taking into consideration the implementation of sub-populations and the migration process from the island model. The results presented were better than canonical version of Cultural Algorithm in most cases.

Island model (Multi Population Genetic Algorithms)

Multi population Genetic Algorithms (MGAs) or Island Model, is an extension of traditional single-population Genetic Algorithms (SGAs) by dividing a population into several subpopulations within which the evolution proceeds and individuals are allowed to migrate from one sub-population to another. Different values for parameters such as selection, recombination and mutation rate can be chosen for each sub-population. Normally, the basic island model uses the same values for these parameters in all sub-populations. In order to control the migration of individuals, several parameters were defined such as:

- the communication topology that defines the connections between sub-populations,

- a migration rate that controls how many individuals migrate, and

- a migration interval that affects the frequency of migration. In addition, migration must include strategies for migrant selection and for their inclusion in their new sub-populations (Aguire, 2000).

The sub-populations size, communication topology (its degree of connectivity), migration rate and migration frequency are important factors related to the performance of distributed GAS. In general, it has been shown that distributed GAs can produce solutions with similar or better quality than single population GAs, while reducing the overall time to completion in a factor that is almost in reciprocal proportion to the number of processors

(Aguire, 2000). In the island model GA, the sub-populations are isolated during selection, breeding and evaluation. Islands typically focus on the evolutionary process within sub-populations before migrating individuals to other islands, or conceptual processors, which also carry out an evolutionary process. At predetermined times, during the search process, islands send and receive migrants to other islands. There are many variations of distributed models, e.g. islands, demes, and niching methods, where each requires numerous parameters to be defined and tuned (Gustafson, 2006). An example of the communication topology, can be defined as a graph in which the subpopulations Pi (i = 0, 1,..., K - 1) are the vertices and each defined edge Li,j specifies a communication link between the incident vertices Pi and Pj (neighbor sub-populations) (Aguire, 2000). In general, assuming a directed graph for each defined link Li,j we can indicate the number of individuals Ri,j that will migrate from P to Pj (migration rate) and the number of generations M between migration events (migration interval). The communication topology and migration rates could be static or dynamic and migration could be asynchronous or synchronous. Various strategies for choosing migrants have been applied. Two strategies often used to select migrants are selection of the best and random selection. For example, the migration can implement a synchronous elitist broadcast strategy occurring every M generation. Each sub-population broadcasts a copy of its R best individuals to all its neighbor subpopulations. Hence, every sub-population in every migration event receives migrants. Figure 5 illustrates a communication topology +1+2 island model in which each sub-population is linked to two neighbors (L = 2). In this example, the sub-population P0 can send individuals only to P1 and P2 and receive migrants only from P4 and P5.

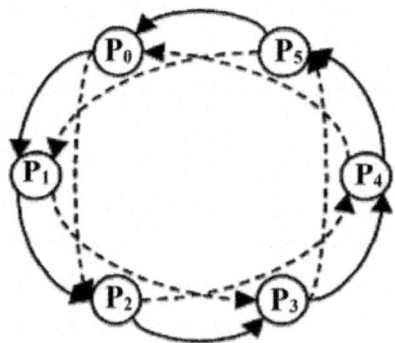

Figure 5: +1+2 communication topology.

CULTURAL ISLAND MODEL (CA-IM)

In this section is presented an approach about the communication topology for migration process implemented in a Cultural Algorithm based on the island model. As noted earlier in the classical island model implementation, there are sub-populations connected with as ring structure. Individuals in classical island model are migrated after every migration-interval (M) among generations and the best-worst migration policy is used. The approach utilized in this work is an adaption and implementation of the island model on the cultural structure here identified as "Cultural Island Model" (CA-IM), briefly introduced in Silva & Oliveira (Silva & Oliveira, 2009). The implementations have become simple because the same CAs structures were used as much the evolutionary structure as the belief space that is the main characteristic present in CAs. The main characteristic present in CA-IM is the link between main belief spaces (from main population) and secondary belief space (from multi population). They store information about independent evolution for main population and sub-populations respectively, i.e. the cultural evolutions occur in parallel among the main population and the sub-populations of the islands. The link of communication between two Belief Spaces, allows migration between the best individuals stored in the cultural knowledge structure implemented. Figure 6 shows the framework correspondent to the proposed structure.

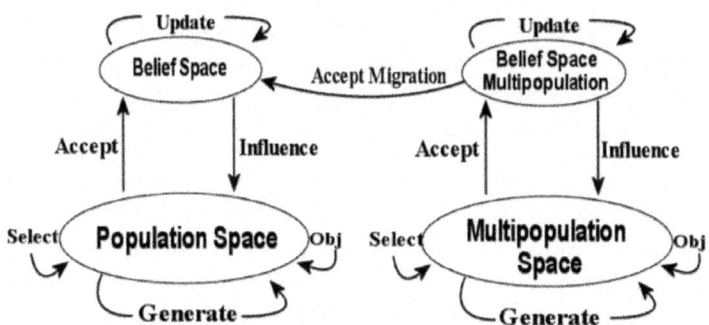

Figure 6: Framework of model proposed

Migrations from islands occur through Belief Space Multipopulation structure that perform the communication process among sub-populations and send the best individuals through Accept Migration. It occurs in a predefined interval whose parameter is M (every M generation) where the best individuals are evaluated by acceptance function and updated in each belief space. The migration from Belief Space Multipopulation to Main Belief Space is implemented as a number of individuals which are considered as a set of exemplars to the rest of the population (Situational Knowledge). It is important to note that CA-IM provides a continuous verification between the last solution (optimum value) found and the current solution. Then, it computes the number of generations where don't occur improvements. Thus, if the distance between the last generation, where the current solution was found, and the current generation is high then the sub-populations are eliminated and recreated randomly. As for CA-IM, there is a fixed difference for this occurrence in the range of 60 to 100 generations. If a new solution is not found in this range, then the sub-populations of the islands (Multipopulation Space) as well as the cultural information about all sub-populations (Belief Space Multipopulation) are recreated randomly by algorithm.

Mutation and recombination

In mutation operation the cultural knowledge (such as situational knowledge) as well as the standard binary mutation operation (known as "bit-flip mutation") is utilized. If the cultural knowledge is utilized during the mutation process, the mutated chromosome genes are replaced by the best genes from chromosome stored in situational knowledge with P_M probability, otherwise, the genes are inverted by bit-flip mutation. The chromosome chosen among a set of chromosomes from situational knowledge can be the best chromosome or a random chromosome. The bit-flip mutation is a common operation applied in evolutionary algorithms to solve a problem with binary representation. Consequently, each bit from current mutated chromosome is flipped, i.e. the value of the chosen gene is inverted also with probability of mutation P_M. Figure 7 shows the pseudo-code of mutation utilized by CA-IM.

In recombination operation the cultural knowledge as well as the standard binary recombination operation (known as "uniform recombination") is also utilized. In the uniform recombination the bits are randomly copied from the first or from the second parent to genes in the offspring chromosomes, in any sequence of ones and zeros. Figure 8 shows the pseudo-code of CA-IM recombination. If the cultural knowledge is utilized during the recombination process, the chromosome genes are replaced by the best genes from chromosome stored in situational knowledge with PR probability. Otherwise, the genes are replaced with genes from their parents. Here only the best chromosome is chosen from situational knowledge during the recombination process.

```
CA-IM  Mutation
1- Get initial Parameters:
    • P_M (Probability of Mutation);          • C (Current Chromossome) ;
    • S [ ] (Situational Knowledge);          • Sbest (best chromosome of S[ ]);
    • Srandom= random chromosome of S[ ];
2- Create C '(Chromossome Mutation)
    If (random<=0.5)
    {
            for (int i=0; i<genotypeLength; i++)
            {
                    if (random <=P_M)
                    {
                            if (random <=0.5)
                                C'[i]=Sbest[i];
                            else
                                    C'[i]=Srandom[i];
                    }
            }
    }
    else
    {
        for (int i=0; i<genotypeLength; i++)
        {
            If (random <=P_M)
                C'= flip(C, i);
        }
    }
3- Return C';
```

Figure 7: Mutation pseudo-code

```
CA-IM  Recombination
1- Get initial Parameters:
    • P_R (Probability of Recombination);
    • C1 (Current Chromosome -1);            // Parents Chromosomes
    • C2 (Current Chromosome -2);            // Parents Chromosomes
    • S [ ] (Situational Knowledge);
    • Sbest=best chromosome of S [ ];
2- Create C 1' and C2' // Offsprings Chromosome Recombination )
    for (int i=0; i<genotypeLength; i++)
    {
        If (random<= P_R)
        {
            If (random <=0.5)
            {
                if (random<= 0.5)
                {
                        C1'[i] = Sbest[i];
                        C2'[i] = Sbest[i];
                }
                else
                {
                        C1'[i] = C2[i];
                        C2'[i] = Sbest[i];
                }
            }
            else
            {
                    C1'[i] = C2[i];
                    C2'[i] = C1[i];
            }
        }
        else
        {
            C1'[i] = C1[i];
            C2'[i] = C2[i];
        }
    }
3- Return Sons (C1' and C2');
```

Figure 8: Recombination pseudo-code.

Experimental results and discussion To evaluate the performance of the proposed algorithm CA-IM, a comparison of various tests with Distributed Genetic Algorithms utilizing the same knapsack problems was carried out. To make a comparison two kinds of algorithms based in Distributed GAs (Aguirre et al., 2000): (i) A Distributed canonical GA (denoted as DGA), and (ii) a Distributed GA-SRM (denoted as DGA-SRM) were utilized. The SRM term means "Self-Reproduction with Mutation", and introduces diversity by means of mutation inducing the appearance of beneficial mutations. For the CA-IM algorithm there are two versions: CA-IM_1 and CA-IM_2. The only difference is that CA-IM_1 has a fixed rate for mutation and recombination, while CA-IM_2 has a random rate for mutation and recombination. The standard CA (CAs) is the Cultural Algorithm with single population.

DGA and DGA-SRM

The DGA works with various 0/1 multiple knapsack problems (NP hard combinatorial) which from previous efforts seem to be fairly difficult for GAs (Aguirre et al., 2000). Those algorithms were evaluated on test problems which are taken from the literature. The problem sizes range from 15 objects to 105 and from 2 to 30 knapsacks and can be found in OR-Library (Beasley, 1990). The knapsack problems are defined by: problem (n, m) where n represents the number of objects and m represents the number of knapsacks. Each knapsack

has a specific capacity as well each object has a specific weight. For example, Weing7 (105, 2) represents a MKP with 105 objects and 2 knapsacks. Every experiment presented here has a similar capacity to the work described in DGA and DGA-SRM (Aguire et al., 2000) such as: population size, number of function evaluations in each run and a total of 100 independent runs. Each run uses a different seed for the random initial population. To improve understanding of DGA and DGA-SRM algorithms, some parameters and symbols are presented:

The maximum size of the population is represented by λ_{total} (fixed in 800);

- The parent and offspring population sizes are represented by μ and λ respectively;
- The parameter K represents the number of sub-populations (partitions). Hence, $\lambda *K= \lambda_{total}$ (maximum=800);
- The parameter M is the number of generations between migration events (migration interval) ;
- The symbol N represents the number of times the global optimum was found in the 100 runs;

- The symbol τ represents a threshold (utilized for control of a normalized mutant's survival ratio).

- The symbol T represents the number of function evaluations in each run;

- Average is the average of the best solutions and Stdev is the standard deviation around Average, respectively;

In DGA and DGA-SRM, each sub-population broadcasts a copy of its R best individuals to all of its neighbor sub-populations. Hence, every sub-population in every migration event receives $\lambda_m = L \times R$ migrants, where L is the number of links. When there is no migration and the sub-populations evolve in total isolation, the values corresponding to such a characteristic are denoted by X in the table. The results for knapsack problem Weing7 for DGA and DGA-SRM is shown in the Table 1 (Aguirre et al., 2000).

Table 1: The best results for Weing7 (105, 2) by DGA and DGA–SRM (λtotal =800; T=8x105)

K	λ_m / λ	DGA						DGA-SRM					
		L R	λ	M	N	Average	Stdev	μ	λ	M	N	Average	Stdev
8	0.10	5 2	100	5	0	**1094423.4**	433.38	50	100	80	63	1095421.44	30.84
8	0.05	5 1	100	5	0	1093284.95	733.24	50	100	100	66	1095423.58	29.84
8	0.01	1 1	100	5	0	1089452.96	1082.41	50	100	80	77	**1095430.51**	26.51
8	X	X	100	X	0	1087385.56	1729.4	50	100	X	60	1095419.80	30.86

According to Table 1 the best value found in Average is equal to 1094423.4, for DGA and 1095430.51 for DGA-SRM. Table 1 also indicates that the DGA-SRM improves the results in relation to DGA. Table 2 shows the results found for others knapsack problems by DGA and DGA-SRM. In order to simplify the results shown in Table 2, the following configuration parameters should be considered: K = 16 sub-populations and μ = 25 (Aguirre et al., 2000).

Table 2: The best results for other problems by DGA and DGA-SRM (λ_{total} = 800; T=4x10⁵)

Problem (n, m)	λ_m / λ	DGA						DGA-SRM (τ =0.35)				
		LR	λ	M	N	Average	Stdev	λ	M	N	Average	Stdev
Petersen6 (39,5)	0.01	1 1	50	5	0	**10506.90**	26.11	50	140	77	10614.82	5.82
Petersen7 (50,5)	0.10	5 1	100	5	0	1093284.95	733.24	50	40	89	16535	5.94
Sento1 (60, 30)	0.10	5 1	100	5	0	1089452.96	1082.41	50	40	98	7771.78	1.54
Sento2 (60, 30)	0.10	5 1	100	5	0	1087385.56	1729.4	50	40	84	8721.32	2.11

CA-IM_1 For the algorithm proposed (CA-IM) various parameters and symbols are also considered such as:

- The parameter P is the size of main population;
- The parameter P_M is the probability of mutation and PR probability of recombination.
- The number of islands is K (number of sub-populations);
- The parameter α is the percentage which defines the size of the population of each island at function of P.
- The sub-population size in each island is SI, since SI = α *P.
- The percentage of best individuals in Situational Knowledge on population space is represented by SK_P and the percentage of best individuals in Situational Knowledge on multi population space is represented by SK_M.
- The parameter M is the number of generations between migration events (migration interval). Here M determines the interval of influence from the islands population through the Situational Knowledge.
- The symbol T represents the number of function evaluations in each run;
- The symbol N represents the number of times the global optimum was found in the 100 runs.
- Average is the average of the best solutions and Stdev is the standard deviation around Average;
- Average of generations is the average of the generations whose best solution was found in each run.

For the tests carried out for CA-IM_1, the selection chosen was tournament, whose value is 3, the mutation rate (PM) is 0.025 and recombination rate (PR) is 0.6. The situational knowledge configurations are: SK_P=0.2 and SK_M=0.5. Table 3 shows the results found by CA-IM_1, whose best value found in Average is 1095445 (the optimal value) and in the Average of Generations is 44.49. All values reached have optimum value. However, if Average of Generations is low in relation to total of generations, then this means that the optimum is found in few generations. As it is shown in Table 3, it is possible to observe that CA-IM outperforms DGA-SRM for any configuration such as the number of sub-populations (islands) and size of subpopulation. Similarly, CA-IM also exhibits higher convergence reliability than DGA-SRM with higher values for N and Average with smaller Stdev. These results show that the CA-IM produces higher performance for all utilized parameters.

Table 3: The best results for Weing7 (105, 2) by CA-IM_1 ($\lambda_{tota}1$ =800 and T=8x10⁵).

P	K	α	SI	M	N	Average of Generations	Average	Stdev
400	8	0.125	50	20	100	52.9	1095445	0.0
400	8	0,125	50	05	100	**44.49**	**1095445**	0.0
100	7	1.0	100	05	100	68.87	1095445	0.0

A new result "Average of Generations" was introduced so as to evaluate other type other type of performance whose value represents the average of generations that the optimum value was found for 100 independent runs for each problem presented. Particularly, it occurs when M is low and K is high (see result for Average of Generations). This means that a larger number of islands with small populations produce better convergence. According to Table 3 the best value found in Average is 1095445 (the optimal value) while the Average of generations is 44.49 that means a low value, considering that 500 generations was utilized in each run which T=4x10⁵. This represents 500 generations with a population size equal to 800 (including all subpopulations). Table 4 shows the results for others MKPs found by algorithm CA-IM_1

Table 4: The best results for other problems by CA-IM_1 (λ_{total} = 800, T=4x10⁵)

Problem (n, m)	P	K	α	SI	M	N	Average of Generations	Average	Stdev.
Petersen6 (39,5)	400	8	0.125	50	20	100	30.22	10618.0	0.0
Petersen6 (39,5)	400	4	0,25	100	05	100	**26.29**	10618.0	0.0
Petersen7 (50,5)	400	8	0.125	50	20	100	78.49	16537.0	0.0
Petersen7 (50,5)	400	4	0,25	100	05	100	**71.51**	16537.0	0.0
Sento1 (60,30)	400	8	0.125	50	20	100	100.21	7772.0	0.0
Sento1 (60,30)	400	4	0,25	100	05	100	**87.44**	7772.0	0.0
Sento2 (60,30)	400	8	0.125	50	20	99	185.19	8721.81	0.099
Sento2 (60,30)	400	4	0,25	100	05	100	**166.12**	87722.0	0.0

Thereby, it is possible to observe that CA-IM_1 outperforms DGA-SRM. Similarly, CA-IM_1 also exhibits higher convergence reliability (higher values of N and Average with smaller Stdev) than DGA-SRM. These results show that the CA-IM_1 is able to find global optimal for MKP, taking into consideration the tests results with 100% success. The problem that presented greater difficulty was Sento2, that presented in some cases optimal values near to 100% such as N=98 and N=99. Even with results of N < 100 they are still better than the results obtained in the chosen benchmarks. In the meantime, the implementation of some adjustments allows CA-IM_1 to reach N=100 for Sento2.

CA-IM_2

For the tests carried out for CA-IM_2 the selection chosen was tournament whose value is 3. The mutation rate (P_M) is a random value in a specific interval: P_M= [0.01, 0.5]. The Recombination rate (PR) is also a random value in an interval: PR= [0.1, 0.99]. The situational knowledge configurations are: SKP=0.2 and SKM=0.5. The CA-IM_2 results are presented in Table 5 that shows the results for Weing7 and in Table 6 that shows the results for others knapsack problems.

Table 5: The best results for Weing7 (105,2) by CA-IM_2 (λ_{total} =800, T=8x10^5)

P	K	α	SI	M	N	Average of Generations	Average	Stdev
400	8	0.125	50	20	100	**70.48**	1095445	0.0
400	8	0,125	50	05	100	72.72	1095445	0.0
100	7	1.0	100	05	100	107.11	1095445	0.0

Table 6: The best results for other problems by CA-IM_2 (λ_{total} = 800, T=4x10^5)

Problem (n, m)	P	K	α	SI	M	N	Average of Generations	Average	Stdev.
Petersen6 (39,5)	400	8	0.125	50	20	100	37.89	10618.0	0.0
Petersen6 (39,5)	400	4	0,25	100	05	100	**33.39**	10618.0	0.0
Petersen7(50,5)	400	8	0.125	50	20	100	81.46	16537.0	0.0
Petersen7(50,5)	400	4	0,25	100	05	100	**74.38**	16537.0	0.0
Sento1(60,30)	400	8	0,25	50	20	98	**112.55**	7771.75	1.7717
Sento1(60,30)	400	4	0,25	100	05	100	126.46	7772.0	0.0
Sento2(60,30)	400	8	0.125	50	20	71	183.35	8720.0	3.7199
Sento2(60,30)	400	4	0,25	100	05	88	**173.53**	8721.38	2.1732

The implementation of random rate for mutation and recombination in CA-IM_2 doesn't produce satisfactory results in comparison to CA-IM_1, as it is shown in Table 6. In addition, the Average of Generations from algorithm CA-IM_2 is greater than CA-IM_1 for all knapsack problems. However, in comparison to CA-IM_1, there are few differences in results for Weing7 as is shown in Table 3 and Table 5.

CA-S (Standard CA)

For CA-S we also utilized the same configuration such as: tournament value=3, P_M= 0.025 and PR = 0.6. The situational knowledge configuration is equal to 0.2 (SK$_p$=0.2). Every experiment presented here also consists of

100 independent runs and each run uses a different seed for the random initial population.

Table 7: The best results for all knapsack problems by CA-S (T=4x10^5).

Problem (, m)	P	N	Average	Stdev.
Petersen6 (39,5)	800	97	10617.58	2.4002
Petersen7 (50,5)	800	81	16533.7	6.8703
Sento1 (60,30)	800	100	7772.0	0.0
Sento2 (60,30)	800	82	8721.14	2.4495
Weing7 (105,2)	800	100	1095445.0	0.0

Table 7 shows the results from standard Cultural Algorithm (CA-S) that utilizes single population. According to results, the CA-S reaches optimum average for 100 runs only for Sento1 and Weing7. However, the results from CA-S for Petersen6, Pertersen7 and Sento2 outperform the results presented by DGA-SRM.

CONCLUSION

This work presented a Cultural Algorithm (CA) with single population (CA-S) and multi population (CA-IM) in order to improve the search performance on MKP. It was observed that CA-S improves the convergence reliability and search speed. However, CA-S is not enough to reach global optimum for most problems presented. Our cultural algorithm implementation with island model (CA-IM_1 and CA-IM_2) allows the migration among islands sub-populations and main population through belief space structures that represent the cultural knowledge available in Cultural Algorithms. The results have shown that the CA-IM_1 is better than CA-IM_2 for the benchmarks selected. The results have also shown that the CA-IM_1 and CA-IM_2 perform the optimum search and reach optimum values equally or above the ones reached by algorithms DGA and DGA-SRM that were chosen for comparison. The positive results obtained, give support the idea that this is a desirable approach for tackling highly constrained NP-complete problems such as the MKP. In addition, it is possible that the hybridization of cultural algorithms based on population of GA with local search techniques improves the results obtained by standard CAs. In a future work, a study will be done about the behavior of the sub-populations that are eliminated and recreated randomly. In addition a local search will be implemented to CAs as much for standard CA (single population) as for CA-IM (multi population) so as to verify improvements on these algorithms.

REFERENCES

1. References Adeyemo, J.A. (2011). Reservoir Operation using Multi-objective Evolutionary AlgorithmsA Review, In: Asian Journal of Scientific Research, Vol.4, No. 1, pp.16-27, February 2011, ISSN 19921454. Agapie, A., Fagarasan, F. & Stanciulescu, B. (1997).

2. A Genetic Algorithm for a Fitting Problem, In: Nuclear Instruments & Methods in Physics Research Section A, Vol. 389, No. 1-2, April 1997, pp. 288-292, ISSN 0168-9002.

3. Aguirre, H. E. & Tanaka, K. (2006). A Model for Parallel Operators in Genetic Algorithms, In: Springer Book Series Studies in Computational Intelligence, Parallel Evolutionary Computations , Nedjah, N., Alba, E. & Macedo M., L., pp.3-31, Springer, ISBN 9783540328391, Berlin Heidelberg.

4. Aguirre, H. E., Tanaka, K., Sugimara, T. & Oshita, S. (2000). Improved Distributed Genetic Algorithm with Cooperative-Competitive Genetic Operators, In: Proc. IEEE Int.

5. Conf. on Systems, Man, and Cybernetics, Vol.5, ISBN 0-7803-6583-6, pp. 3816-3822, Nashville, TN, USA, October 8-11 2000. Aguirre, H. E., Tanaka, K. & Sugimura, T. (1999).

6. Cooperative Model for Genetic Operators to Improve GAs In: International Conference on Information Intelligence and Systems, ISBN 0-7695-0446-9, pp. 98–106, Bethesda, MD, USA, 31 Oct. - 03 Nov., 1999. Ahmed, Z. & Younas I. (2011).

7. A Dynamic Programming based GA for 0-1 Modified Knapsack Problem, In: International Journal of Computer Applications, Vol. 16, No.7, February, 2011, pp. 1–6, ISSN 09758887. Back, T., Fogel D. B. & Michalewicz Z., (Ed(s).). (1997).

8. Handbook of Evolutionary Computation, Oxford University Press, ISBN 0-7503-0392-1, UK. Beasley, J. E. (1990). Multi- Dimensional Knapsack Problems, In: OR-library, Date of Access: September 2011, Available from: http://people.brunel.ac.uk/~mastjjb/jeb/orlib/ mknapinfo.html.

9. Braun, H. (1991). On Solving Traveling Salesman Problems by Genetic Algorithms, In: Parallel Problem Solving from Nature—Proceedings of 1st Workshop, Vol. 496 of Lecture Notes in Computer Science, H.P. Schwefel and R. Manner, Vol. 496, pp. 129-133, Springer, ISBN 3-540-54148-9, Dortmund, FRG, October 1-3 1990.

10. Chu, P. C. & Beasley J. E. (1998). A Genetic Algorithm for the Multidimensional Knapsack Problem, In: Journal of Heuristics, vol. 4, no. 1, June 1998, pp. 63–86, ISSN:1381-1231.

11. Eiben, A. E. & Smith, J.E. (2008). Introduction to Evolutionary Computing, Springer, Second edition, ISBN 978-3-540-40184-1, Amsterdam, NL. Fukunaga, A. S. & Tazoe, S. (2009).

12. Combining Multiple Representations in a Genetic Algorithm for the Multiple Knapsack Problem, In: IEEE Congress on Evolutionary Computation, ISBN 978-1-4244-2958-5, pp. 2423 – 2430,

13. Trondheim, 18-21 May, 2009. Gallardo, J. E., Cotta C. & Fernández A. J. (2007). On the Hybridization of Memetic Algorithms With Branch-and-Bound Techniques, In: IEEE

14. Transactions on Systems, Man, and Cybernetics, Part B, Vol. 37, No. 1, February 2007, pp. 77-83, ISSN 1083- 4419. Glover, F. & Kochenberger, G. A. (1996).

15. Critical Event Tabu Search for Multidimensional Knapsack Problems, In: Meta-Heuristics: Theory and Applications, Osman, I.H. & Kelly, J.P., pp. 407-427, Springer, ISBN 978-0-7923-9700-7,

16. Boston, USA. Gunther, R. R. (1998). An Improved Genetic Algorithm for the Multiconstrained 0–1 Knapsack Problem, In: Evolutionary Computation Proceedings. IEEE World Congress on Computational Intelligence, ISBN 0-7803-4869-9, pp.207-211, Anchorage, AK, May 4-9 1998.

17. Gustafson, S. & Burke, E.K. (2006). The Speciating Island Model: An Alternative Parallel Evolutionary Algorithm, In: Journal of Parallel and Distributed Computing, Vol. 66, No. 8, August 2006, pp. 1025-1036, ISSN 07437315.

18. Iacoban, R., Reynolds, R. & Brewster, J. (2003). Cultural Swarms: Modeling the Impact of Culture on Social Interaction and Problem Solving, In: IEEE Swarm Intelligence Symposium. ISBN 0-7803-7914-4, pp. 205–211, University Place Hotel, Indianapolis, Indiana, USA,

19. April 24-26 2003. Jin, X., & Reynold, R. G. (1999). Using Knowledge-Based System with Hierarchical Architecture to Guide the Search of Evolutionary Computation, In: Proceedings of the 11th IEEE International Conference on Tools with Artificial Intelligence, ISBN 0-7695- 0456-6, pp. 29–36, Chicago, Illinois, November 08-10 1999.

20. Ku, S. & Lee, B. (2001). A Set-Oriented Genetic Algorithm and the Knapsack Problem, In: Proceedings of the Congress on Evolutionary Computation, ISBN 0-7803-6657-3, pp. 650– 654, Seoul, South Korea, May 27-30 2001. Kumar, R. & Singh, P. K. (2010). Assessing Solution Quality of Biobjective 0-1 Knapsack Problem using Evolutionary and

Heuristic Algorithms, In: Applied Soft Computing, Vol. 10, No 3, June 2010, pp. 711 – 718, ISSN 1568-4946.

21. Martello, S. & Toth, P. (1990). Algorithms and Computer Implementations, John Wiley & Sons, ISBN 0471924202, New York. Nowostawski , M. & Poli , R. (1999).

22. Parallel Genetic Algorithm Taxonomy, In: Proceedings of the Third International conference on knowledge-based intelligent information engineering systems, ISBN 0780355784, pp. 88-92,

23. Adelaide, August 1999. Reynolds, R. G., & Peng, B. (2004). Cultural Algorithms: Computational Modeling of How Cultures Learn to Solve Problems,

24. In: Seventeenth European Meeting on Cybernetics and Systems Research, ISBN 3-85206-169-5, Vienna, Austria, April 13-16 2004. Reynolds, R. G. & Chung C. (1996).

25. The Use of Cultural Algorithms to Evolve Multiagent Cooperation, In: Proc. Micro-Robot World Cup Soccer Tournament, pp. 53–56. Taejon, Korea, 1996. Reynolds, R. G. (1994),

26. An Introduction to Cultural Algorithms, In: Proceedings of the Third Annual Conference on Evolutionary Programming, ISBN 9810218109, pp. 131-139, San Diego, California, February 24-26 1994.

27. Reynolds, R. G. (1999). Chapter Twenty-Four; Cultural Algorithms: Theory and Applications, In: New Ideas in Optimization, Corne, D., Dorigo, M. & Glover F., pp. 367-377,

28. McGraw-Hill Ltd., ISBN 0-07-709506-5, UK, England. Silva, D. J. A. & Oliveira R. C. L. (2009). A Multipopulation Cultural Algorithm Based on Genetic Algorithm for the MKP, In: Proc. of the 11th Annual conference on Genetic and evolutionary computation, ISBN 978-1-60558-325-9, pp. 1815-1816,

29. Montreal, Québec, Canada, July 8-12 2009. Sivanandam, S.N. & Deepa, S. N. (2007). Introduction to Genetic Algorithms, (1st), Springer, ISBN 978-3-540-73189-4, New York.

30. Sivaraj, R. & Ravichandran,T. (2011). An Improved Clustering Based Genetic Algorithm for Solving Complex NP Problems, In: Journal of Computer Science, Vol. 7, No. 7, May 2011, pp. 1033-1037, ISSN 15493636. Spillman, R. (1995).

31. Solving Large Knapsack Problems with a Genetic Algorithm, In: IEEE International Conference on Systems, Man and Cybernetics, Vol. 1, ISBN 0-7803-2559-1, pp 632 -637,

32. Vancouver, BC, Canada, October 22-25 1995. Tavares, J., Pereira, F. B. & Costa, E. (2008). Multidimensional Knapsack Problem:

33. A Fitness Landscape Analysis, In: IEEE Transactions on Systems, Man and Cybernetics, Part B: Cybernetics, Vol. 38, No. 3, June 2008, pp.604-616, ISSN 1083-4419.

34. Tomassini, Marco (2005). Spatially Structured Evolutionary Algorithms: Artificial Evolution, Space and Tim - Natural Computing Series (1st), Springer, New York, Inc., ISBN 3540241930,

35. Secaucus, NJ, USA. Zoheir, E. (2002). Solving the 0/1 knapsack Problem Using an Adaptive Genetic Algorithm, In: Artificial Intelligence for Engineering Design, Analysis and Manufacturing, Vol.16, No. 1, January 2002, pp.23-30, ISSN 08900604.

CITATION

CHAPTER 1

C. Nataraj, A. Jalali and P. Ghorbanian (2012). Application of Computational Intelligence Techniques for Cardiovascular Diagnostics, The Cardiovascular System - Physiology, Diagnostics and Clinical Implications, Dr. David Gaze (Ed.), ISBN: 978-953-51-0534-3, InTech, DOI: 10.5772/38032.

CHAPTER 2

Young-Doo Kwon and Dae-Suep Lee (2012). The Successive Zooming Genetic Algorithm and Its Applications, Bio-Inspired Computational Algorithms and Their Applications, Dr. Shangce Gao (Ed.), ISBN: 978-953-51-0214-4, InTech, DOI: 10.5772/36400.

CHAPTER 3

Eduardo Fernández-González, Inés Vega-López and Jorge Navarro-Castillo (2012). Public Portfolio Selection Combining Genetic Algorithms and Mathematical Decision Analysis, Bio-Inspired Computational Algorithms and Their Applications, Dr. Shangce Gao (Ed.), ISBN: 978-953-51-0214-4, DOI: 10.5772/38121

CHAPTER 4

Annibal Hetem Jr. (2012). The Search for Parameters and Solutions: Applying Genetic Algorithms on Astronomy and Engineering, Bio-Inspired Computational Algorithms and Their Applications, Dr. Shangce Gao (Ed.), ISBN: 978-953-51-0214-4, InTech, DOI: 10.5772/38002.

CHAPTER 5

Sertan Erkanli, Jiang Li and Ender Oguslu (2012). Fusion of Visual and Thermal Images Using Genetic Algorithms, Bio-Inspired Computational Algorithms and Their Applications, Dr. Shangce Gao (Ed.), ISBN: 978-953-51-0214-4, InTech, DOI: 10.5772/36139.

CHAPTER 6

Julio César Martínez-Romo, Francisco Javier Luna-Rosas, Miguel Mora-González, Carlos Alejandro de Luna-Ortega and Valentín López-Rivas (2012). Optimal Feature Generation with Genetic Algorithms and FLDR in a Restricted-Vocabulary Speech Recognition System, Bio-Inspired Computational Algorithms and Their Applications, Dr. Shangce Gao (Ed.), ISBN: 978-953-51-0214-4, , DOI: 10.5772/36135.

CHAPTER 7

Kim Soon Gan, Patricia Anthony, Jason Teo and Kim On Chin (2012). Performance of Varying Genetic Algorithm Techniques in Online Auction, Bio-Inspired Computational Algorithms and Their Applications, Dr. Shangce Gao (Ed.), ISBN: 978-953-51-0214-4, InTech, DOI: 10.5772/36758.

CHAPTER 8

Pedro Rocha, Alexandre Pigozzo, Bárbara Quintela, Gilson Macedo, Rodrigo Santos and Marcelo Lobosco (2012). Modelling the Innate Immune System, Bio-Inspired Computational Algorithms and Their Applications, Dr. Shangce Gao (Ed.), ISBN: 978-953-51-0214-4, InTech, DOI: 10.5772/38690.

CHAPTER 9

K. A. Folly and S. P. Sheetekela (2012). Optimal Design of Power System Controller Using Breeder Genetic Algorithm, Bio-Inspired Computational Algorithms and Their Applications, Dr. Shangce Gao (Ed.), ISBN: 978-953-51-0214-4, InTech, DOI: 10.5772/38447

CHAPTER 10

Deam James Azevedo da Silva, Otávio Noura Teixeira and Roberto Célio Limão de Oliveira (2012). Performance Study of Cultural Algorithms Based on Genetic Algorithm with Single and Multi Population for the MKP, Bio-Inspired Computational Algorithms and Their Applications, Dr. Shangce Gao (Ed.), ISBN: 978-953-51-0214-4, InTech, DOI: 10.5772/36366.

INDEX

A

Abundances 90, 112
Artificial neural networks (ANN) 3
Astrophysics 85

B

Bandwidth (BW) 169
Biometric technologies 115
Blood pressure (BP) 1, 19, 31
Breeder Genetic Algorithm (BGA) 242, 243
Brightening 117

C

Calculation 90
Coefficients 147, 151, 152, 153, 154, 155, 156, 168, 172, 174, 175, 176, 177
Con Figureurable 197
Continuous Genetic algorithms (CGA). 119
Continuous wavelet transform (CWT) 6
Conventional PSS (CPSS) 243, 248
Conventional PSS (CPSS). 243
Cultural Algorithm 260, 266, 269, 271, 274, 279, 282
Cultural Algorithms (CAs) 259, 266

D

Decision Makers (DM), 58
Differential Evolution, Immune Systems, 260
Drawback 241
Dynamic range compression (DRC 126
Dynamic Time Warping (DTW) 156

E

Electric Power Distribution Protective Devices 106, 111
Electrocardiogram (ECG) 2
Enhancement Technique for Nonuni- form and Uniform-Dark Images (ETNUD) 127
Enhancement Technique for Nonuni- form and Uniform-Dark Images (ETNUD). 127
European Southern Observatory (ESO) 92
Evolutionary algorithms (EAs) 261
Evolutionary Computation (EC) 259

F

Fernandez 57, 58, 59, 68, 71, 72, 75, 80, 81
Fieldprogrammable gate array-based (FPGA) 5

Fisher Discriminant Ratio (FDR) 149
Fisher's linear discriminant ratio
    (FLDR). 169
Fisher's Linear Discriminant Ratio
    (FLDR) 176
Fused image (FI). 132
Fuzzy clustering neural network archi-
    tecture (FCNN) 5

G

General-purpose 218
Genetic algorithm 181, 184, 194
Genetic Algorithm 116, 118, 119, 133,
    134, 141, 142, 144, 146
Genetic Algorithm (GA), 259, 263
Genetic algorithms (GA) 133
Genetic Algorithms (GAs 241, 243
Genetic Algorithms (GAs) 83, 241, 243,
    261
Global limitations 158
Glover & Kochenberger 260
Graphics Processing Units (GPGPUs)
    218
Grefenstte, Schaffer 193
Gregorio-Hetem 85, 86, 87, 88, 89, 90,
    92, 93, 94, 99, 112, 113

H

Harrington pump 95, 96, 99
Harrington Pumps 94, 95, 111
Heart rate (HR) 1, 19
Heuristic 68, 76
Hidden Markov Models HHM 168
Human Immune System (His) 217
Human visual system (HVS) 126

I

Illuminance Reflectance Model for En-
    hancement (IRME) 126
infrared (IR) 143
Infrared (IR) 115, 116
Instantaneous field-of-view (IFOV), 132

L

Linear Predictive Coding (LPC) 152
Linear Predictive Coding, or LPC 151
Lipopolysaccharide (LPS) 217

M

Metallicities 90, 112
Modern intensive care units (ICU) 1
Multi Choice Knapsack Problems
    (MCKP) 261
Multi-criteria approaches are MAUT 58
Multi-layered perceptron (MLP) 5
Multiple Knapsack Problems (MKP)
    261
Multi Population 260, 261, 269, 286
Multi-Scale Retinex (MSR) 126
Multi-Scale Retinex with Color Restora-
    tion (MSRCR) 126

N

Neighborhood-Dependent Approach for
    Nonlinear Enhancement (AIN-
    DANE) 127
Network composed 217
Non-linear radiative-density 85
NP-complete problems 85

O

operating condition 246, 248

P

Parallel Genetic Algorithms (PGAs) 268
Parameterized function 117
Parametrization 94, 95, 111
Parseval's theorem 148, 161, 162, 170,
    177
Partial Differential Equations (PDEs)
    217, 220
Pixel-level image fusion (PLIF) 133
Principal component analysis (PCA) 4,
    7

R

Random numbers 124, 135
Receiver operating curve (ROC), 149
Representation-Switching GA 262
Restricted-vocabulary 167
Rumelhart 163

S

SAIDI (System Average Interruption
    Duration Index 106
SAIFI (System Average Interruption
    Frequency Index) 107
Scale of observation. Another important
    distinction from the STFT 6
Self adaptive crossover and mutation
    (SACM) 206
self-adaptive crossover (SAC) 206
self –adaptive mutation schemes (SAM)
    206

Separability measuring criterion 149,
    151
Sequential Backward Selection 150
Single-Scale Retinex (SSR), 126
Sivanandam 268, 282
Spectral energy distribution (SED 85
standard Distributed GA (DGA) 262
Support vector machines (SVM) 149
Support vector machine (SVM), 5

T

Temporal evolution 226, 227, 228, 229,
    230, 231, 232, 233, 234, 235, 236
Thermal IR imagery 116

W

Wavelet transform (DWT) 4
Wavelet transforms (WT) 3